THE MAN WHO TALKS TO WHALES

Also by Jim Nollman

---

*Why We Garden*
*The Beluga Café*
*Spiritual Ecology*
*The Charged Border*

# The Man Who Talks to Whales

Jim Nollman

Sentient Publications, LLC

Printed in the United States of America.

*Illustrations by Jim Nollman*
*Cover design by Ana Dayva*
*Cover illustration by Surina Emily Ebsen*
*Author photograph by Ron Rabin*
*Book design by Paul Kobelski*

LIBRARY OF CONGRESS CATALOGING-IN-PUBLICATION DATA

Nollman, Jim.
The man who talks to whales : the art of interspecies communication / Jim Nollman.
p. cm.
Rev. ed. of: Animal dreaming. c1987.
Includes bibliographical references.
ISBN 0-9710786-2-9
1. Human-animal communication. I. Nollman, Jim. Animal dreaming. II. Title.
QL776 .N64 2002
591.59—dc21

2001006102

SENTIENT PUBLICATIONS
A Limited Liability Company
1113 Spruce Street
Boulder, CO 80302
www.sentientpublications.com

*To Katy, who helps me keep at it*

# *Contents*

# Preface

If you ask an Australian aborigine what his "dreaming" is, he will most likely give you the name of some animal, a feature of the landscape, a plant, or a constellation. This is his origin, the place from where his Spirit came. The Dreamtime pertains to the stories and creation myths that accompany this relationship between a person, his totem, and the environment. The book *People of the Dreamtime,* an Australian account of the life and times of these aboriginal people, describes it so:

> Until recent times these stories constituted a total system of belief for the aboriginal people—an explanation of the universe, of the tribal territories, and of the animate and inanimate features of the countryside, a validation and reinforcement of the workings of aboriginal society, a book of rules for conducting normal and abnormal circumstances, a promise of the continuance of life for one's children and one's children's children.

This book, *The Man Who Talks to Whales,* also describes a relationship between a person, myself, and an environment, in this case, both natural and human. You will also meet several other characters, both human and nonhuman; all sentient, all original, and all with stories of their own. These include dolphins, seagulls, mosquitoes, buffalo, bears, and mitochondria, to name just a few. I have taken that old concept of the Dreamtime and its "promise of the continuance of life" very much to heart in this writing. If we can properly dream our future, then certainly it has a much better chance of becoming reality.

# Acknowledgments

TO LIST ALL THE PEOPLE who were instrumental in supporting this work would be a monumental work in itself. In lieu of that, I say that you know who you are, and thank you. Beyond that, six people stand out, and I affectionately acknowledge each of them:

Gigi Coyle, Sandra Wilson, and Alan Slifka, for believing in this work of interspecies communication; Martin Marix Evans, for taking a very raw manuscript and forcing me to rewrite it so that it was worthy of publication; and Frank Robson, who for years has been my main mentor in this business of learning to talk to animals. And my wife Katy, who has long served as my guide to happy human social relations.

I also acknowledge the orcas, who continue to show me that the experience of the Dreamtime is real and very much of this world.

Parts of this book have been published, in slightly different form, in *Omni*, *New Age Journal*, *CoEvolution Quarterly*, *The Whalewatcher*, *Magical Blend*, and *The Greenpeace Chronicle*.

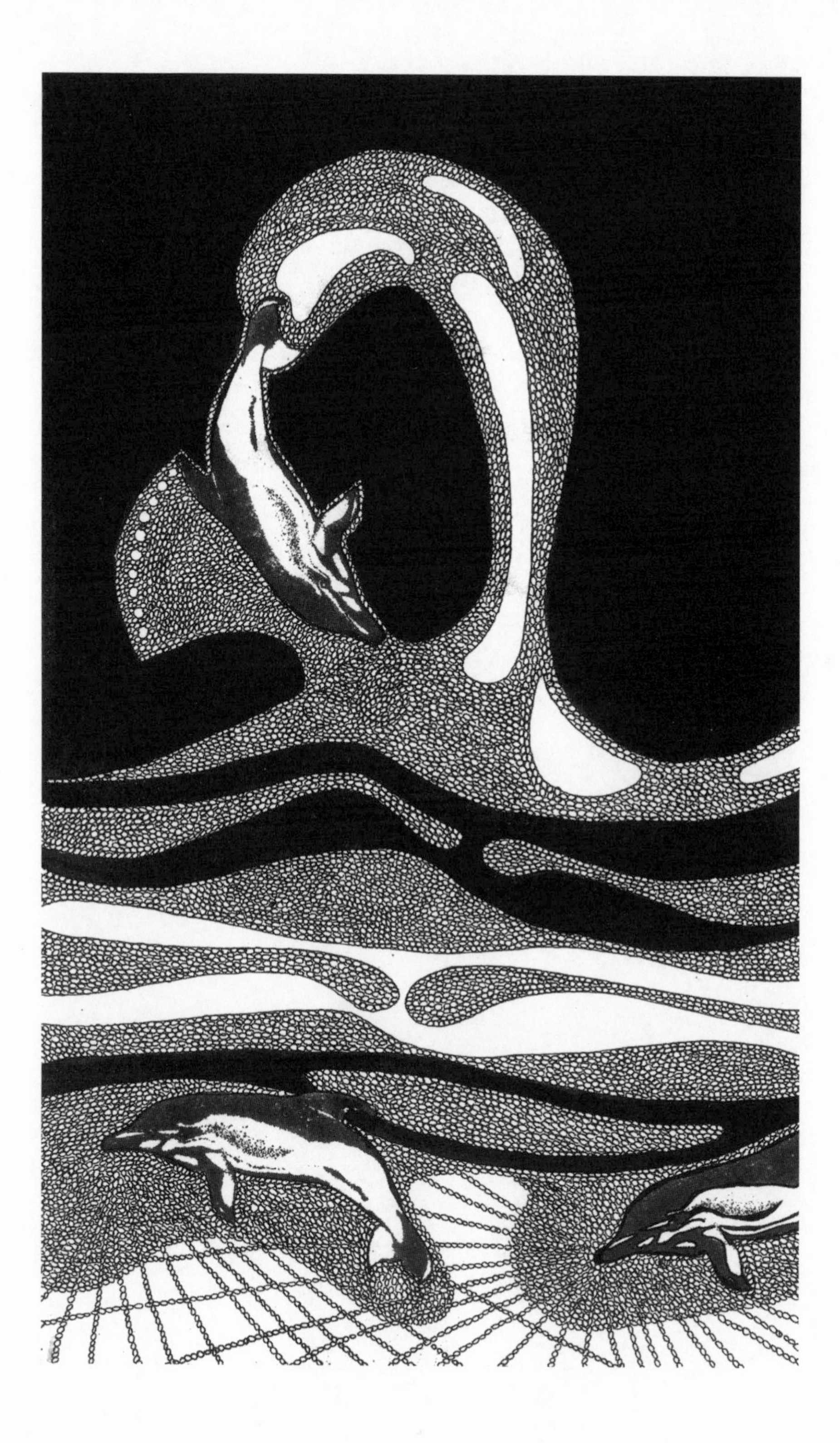

# Introduction

SWIMMING STRAIGHT OUT into the ocean, two hundred yards from shore on this becalmed, coral-studded, blue-green sea. I'm not much of a long-distance swimmer, so I feel only too happy supporting my weight by cradling a stainless steel and brass, hollow sound sphere known as a waterphone. The waterphone, composed of vacuum cleaner tube, pizza plate, and welded salad bowls, is a musical instrument.

Directly in front, twenty or more fins roll out over the ocean surface in close formation. These are spinner dolphins. All the accounts describe them as highly intelligent as well as friendly and playful. But still, this is the first time that I have ventured so far from shore, all alone, with so many large sea creatures about. One small part of my media-cluttered brain brings into focus various shark-infested movie set oceans. These are not sharks, I repeat once more. These are dolphins, known for saving drowning swimmers. Dolphins perceive their environment through the use of sonar—through their ears. They might like music. They might even try to communicate with a human if the tune is just right. No, not in English. I don't know how. I don't really know what I'm trying to say, either to myself or to the dolphins. Communicate what?

I stroke the main tube of the waterphone. When the metallic surface is caught just right, the whole sphere vibrates. Cradling the instrument properly, this vibration courses clear through my sonically transparent body and so into the water. The process tickles.

But the spinners seem totally disinterested in this sound as they continue about their business of traveling into the middle of the bay. I stop rubbing the waterphone and stick my head into the water to listen. I hear nothing but the low bellowing of the waterphone.

So I alter my technique to a shorter, more rhythmical tune by rubbing a palm directly across the many prongs that rim the equator of the sphere. Now the beat is clear and simple: five seconds of sound, and then five seconds of silence. Once again I stick my head in the water. But now the sphere is vibrating so sharply that it actually hurts my ears. I am told that sound is five times more intense underwater than above. Does it hurt the dolphins' ears as well?

The spinners keep their distance. So I change my tune a third time. By striking the center tube with a mallet while simultaneously immersing various sections of the sphere into the water, a ringing tone is produced that rises and falls dramatically in pitch. Cartoon music: *WWWwwggGGG.*

Immediately I sense a change. The dolphins turn from their straight and narrow path and swim right at me. Then, fifty yards away, they form into a tight, precisely patterned circle. Now they turn and turn on their axis for a minute or more. The motion reminds me of nothing so much as Israeli folk dancing.

One at a time the dolphins break from the circle. They move in very close: forty, thirty-five, twenty-five feet away. It is so difficult to judge distance accurately with my eyes a mere six inches off the water's surface. Now their eyes are quite visible as several of their number lift their heads high above the water's surface to examine this unusual vibrating being who swims in their midst. They are about the same size as me. For an instant, the quality of their looking makes me think of human beings who are dressed up in dolphin suits. Mostly, however, I am struck dumb by the immense power of their breathing. It is champagne corks popping all over the ocean.

And all the time I continue to draw these long sliding notes from the throat of the waterphone. The sphere sounds so clear and fruity: now like a church organ, now like an oriental gong with the hiccups. I dunk my head and open my eyes. Seven blurry figures scoot past the edge of vision. The dolphins seem so much closer from underwater. I listen for twenty seconds, ascend for a deep breath of air, descend again. Up and down, over and over again. The technique is somewhat frustrating in its clumsiness. This human is inadequately built for the task of communing with these creatures of the sea. And I still hear nothing but the glissando of the waterphone. I had been led to believe that dolphins vocalize nearly all the time. Perhaps they are vocalizing, but at a frequency much higher than my ears can fathom. And then I hear it. It seems so far away, like the song of elves from the heart of the forest; like the whispering of children from the opposite end of a cathedral. As in the hearing tests we take as youngsters, the sound vanishes as soon as I concentrate on anything else. Unfortunately, I must concentrate most of my energy on the fundamental necessity of drawing breath.

But if I feel inadequate to the task of communicating with the dolphins, the dolphins themselves do not seem to notice. One of them darts directly beneath my feet. I'm still smiling; that's a good sign. I have heard so many stories and fables about the psychic abilities of dolphins. Maybe there is something to it. Some people believe the animals can enter a person's head at will, swim around inside, and then leave you with a distinct tingly feeling. Like when you drink ginger ale too fast.

All of us are quite together now—one human and twenty spinners sharing one little corner of the galaxy. And now, so suddenly as to make my blood rush, one of the dolphins jumps six feet clear of the water, and just a few feet away. A few flukes away. In another moment, they all begin jumping clear of the water, spinning and somersaulting about. That one must have leaped fifteen feet straight up before spinning twice, somersaulting once, and then reentering the water with nary a splash. Bravo! All I can manage to do is keep my head above water and watch them with a big foolish grin across my face.

And from the shore, so very far away, an audience of human beings has gathered to watch this free-form interspecies theater. And now, they too are all jumping about, laughing and clapping, slapping each other on the backs, breaking out the cold beer. Everyone everywhere is carrying on like a group of silly children.

I have been a musician most of my adult life. It is only too plain to me why dolphins do so well as entertainers. They are born performers. Someone on the shore blares out a charge call on a trumpet. Someone thumps on a conga drum. And so, the next series of dolphin jumps and spins seems just that much more animated. The people on the shore cheer louder. The dolphins jump higher. What more could any musician ask from a performance? Maybe dolphins should be the totem animal of human musicians, gymnasts, and clowns.

The animals frolic about for no more than ten more minutes. Then they move off, and once again form into their precise, tightly knit circle, turning and turning about. They break from the circle one at a time and are gone.

This human being suddenly feels freezing cold in twenty-five feet of ocean water, much too far from shore for his own comfort. Kicking with his rubber frog's fins, he straddles the waterphone and slowly undulates back to shore. There, other humans stand, ready to talk to him about what has happened.

# 1

# *The Turkey Trot*

Now in the silence the hummingbird hovers
The human being senses the sound
Two ears four ears hear the droning
The drone, high high high on the mountain

High on the mountain the hummingbird hovers
The human being senses the moment
Two eyes four eyes seeking color
The flower, high high high on the mountain

Inside the flower the hummingbird hovers
The human being senses emotion
Now the bird comes outside the flower
The human being senses a change
So changes the universe

High on the mountain the hummingbird flies away
The human being senses the silence.

AT FIVE YEARS OF AGE I was very sure that deer used their antlers as TV antenna. Just as I was able to rotate my own aerial to change the reception, so the deer (or the antelope or the buffalo) could tune in to a particular show by simply turning this way or that. The show itself was projected directly onto the mind's eye of the creature. I wondered if they watched and enjoyed the same shows that I did. I also considered that all these animals might have their own shows: a moose show customized for moose antler reception, a cow show available only to cows. And of course, no matter what the species, each and every one of these shows was broadcast in English. In those formative years, language *meant* English. How did I know this as a fact? I'd seen *Bambi,* that's how.

Things were so straightforward at age five. All the animals possessed the same basic values and intelligence as myself, my family, and especially those other families up there on the TV screen. When we played animals, it was never a lengthy shamanistic process of "taking on" the attributes of that animal. Whether it meant bear, lion, or even beaver or deer, all of us kids let out a few cherished roars, and then chased each other around the neighborhood just exactly the same way we did when we played cops and robbers, or World War II. The animals kept no secrets from me. I even invented a pat explanation for that otherwise tricky matter of incomprehensible bird songs, which were, truly, the only wild animal utterances that I heard with any regularity. It was all just a secret code, just about the same as our Pig Latin. And what did the birds talk *about?* Naturally, they talked about the same things that we all talk about.

Then came Babel: school, growth, experience. And the Lord said, ". . . let us go down, and there confound their language, that they may not understand one another's speech." They say that a human fetus retraces the path of evolution in its development from one-celled creature to human being. This is known as ontogeny recapitulating phylogeny. I say that the process continues long after birth, but now on the level of culture. A five-year-old is much closer to his animal roots than a six-year-old. And so, with each passing year the chasm between us humans and the rest of nature grows wider and wider. One day we wake up to realize that the gap has finally grown too wide ever to bridge it again. For me, the realization came at age ten. I had found a beautiful little ring-necked snake out in the woods and brought it home to live in a shoe box down in the cellar. Later, after the snake turned up in my mother's tumble-dried towels, it took no more than a single well-placed look from her to know for all time that I was not allowed to keep snakes in the house. A year earlier the same message would have taken much explanation and tears. Now I understood: animals are not members of my immediate family. I was on the verge of becoming a responsible member of twentieth-century society.

By age sixteen I was devouring every book that I could find about the animals. Of course, by now I knew that Walt Disney had been wrong all along; his cartoon movies smacked of my recently discovered word, "anthropomorphism." If zoology were a religion, then anthropomorphism would be its mortal sin. The dictionary defines it as attributing human characteristics and emotions to animals. Dogs cannot laugh, dolphins do not smile, and deer do not possess ingrown TV antennas.

But by sixteen, I had also begun that curious and sometimes painful human process of learning to think for myself. I found myself agreeing with the prevailing sentiments of the natural sciences only up to a point. What I disagreed with was the pedestal from which the zoologists and behavioralists looked down upon the animals. And even then, it seemed important to me. After all, the sentiments that they felt, the methods that they employed, were the same sentiments and methods that I and every other sixteen-year-old were being taught in school. Certainly deer do not converse in English. But then, dogs do sometimes express laughter by wagging their tails. And it may be true that a dolphin's smile is a result of the set of the musculature of the mouth. Yet still, I wanted to know why the dolphins have so much to smile about. And if laboratory experiments could verify that animals become neurotic under stress (it all seemed so obvious), then why could not the same scientists accept the fact that animals were happy when roaming free? But the worst thing of all was the laboratory experiments themselves. How could any semblance of new "knowledge" justify the cruelty that was being perpetrated on animals in the name of medicine, or science, or any other human endeavor for that matter? Dissection was what we learned in high school biology. And if I complained to the teacher that I would not, under any circumstances, stick a pin into the brain of a leopard frog, she, and most of the "serious" students, looked upon me as squeamish, or even a coward.

I experienced animals from a place of deep respect, something not easily verbalized at sixteen, but believed in nonetheless. Furthermore, I could have easily imagined spending the rest of my life dedicated to working with them. But the animals that my culture taught me about were not the same animals that I had come to know; and likewise, the people who worked with animals were not doing the kinds of studies that made me want to follow in their footsteps. Instead, I was always left with the bewildering impression that humans thought that animals were some kind of biological *machine,* devoid of emotions, intellect, and independence. And to my impressionable sixteen-year-old mind, a zoologist was a person who captured animals, or who kept animals at a zoo, or tortured them in laboratories. And all these jobs were created to help humans accumulate more information about animals. This information was, somehow, meant to "help" us humans to grow. It helped *only* us humans. To me, animals had to be more than resources and specimens. I did not necessarily want to learn *about* them, so much as I wanted to learn *from* them.

What I did not know at age sixteen was that many other young animal lovers were experiencing exactly the same confusion. At the time, I chose to blame it all on science. But blaming science for a certain human viewpoint is more than a little like blaming history for the fact that humans make war. Science is measurement, a systematized knowledge derived from observation. It was not science that was to blame, but rather the prevailing view of the working scientists of that time. And to unravel the roots of this dead and thus deadly view of animals involves a long excursion through the maze of our recent cultural history. In its totality, it is a journey far beyond the scope of this book. Read Barry Commoner's *The Closing Circle* for one concise account. However, some of the entanglements along the path of that longer excursion are described here—especially those sentiments that have served to impede all of us from accepting the fact of animal communication and wisdom.

Unlike many other young animal lovers of that time, I did not stick to it and, thus, I did not help to change the "dead view" by becoming a zoologist myself. Instead I turned to music. At age sixteen, I became a professional musician by driving the two hundred miles south from the suburbs of Boston to the coffeehouses of New York City to perform folk music in front of an audience. And for the moment, I had found my niche.

Then came college, a formal training in music for theater, and finally, at age twenty-two, a turn toward the big sound of rock and roll. Unfortunately, one day, while performing in a smoky club in San Francisco, it finally dawned on me that there was no glorious future awaiting me in the rock and roll business. The reason was obvious, but it took my training in music for theater for me to permit myself to believe it. You see, when I was not performing in clubs, I never visited them. I did not smoke, rarely drank more than a glass of wine at dinner. Yet here I stood, out in front of a couple of hundred people, all of whom paid their money to enter this same club to have a good time by melding into the general ambience. In other words, I was promoting a lifestyle of which I was not really a part. My rock and roll dream had a built-in flaw. But then, I wanted to know, how else does a musician find work within our culture?

I turned back to the theater, and spent several key years providing music for various dance, pantomime, circus, and drama troupes. Composer John Cage became a major influence. His music assured me that art is, in reality, whatever you can get away with. Once, in just that spirit, I co-produced a media event on a San Francisco beach. The piece deserves retelling, if for no other reason than the fact that it was my first incursion into music designed for the outdoors.

Twelve contact microphones were placed all over the body of an upright piano. Each microphone was hooked up to its own very powerful amplifier. Twelve large speaker columns surrounded the piano, creating a vision from Stonehenge. Then, we soaked the piano with kerosene. At that time in my life I had been suffering from bronchial asthma, a condition brought on whenever I physically overexerted myself. Luckily, I had also learned to cure an attack by sitting quietly, repeating a precise breathing exercise through a bamboo flute. Five hundred people showed up for the event, including one of the national TV news shows. I started a mile down the beach from the piano, lit a kerosene-soaked torch, and then proceeded to sprint along the surf as the sun slowly sank into the sea. By the time that I reached the piano, I could no longer breathe properly. I touched the torch to the piano, which at first burned slowly, and then finally, quite ferociously. Then I sat on the sand between burning piano and the ocean, and commenced to play the "asthma remedy" through a thirteenth microphone set up for the occasion. Over the next full hour, five hundred people watched and listened to the breathy flute sonata accompanied by the very loud snap, crackle, and pop of the spectacularly burning piano.

The upshot was that I emerged from the event as something of an authority in the poorly understood field of music and health.

*Asthma Burning* was a prime example of the conceptual music so popular in the Bay Area at that time. Yet although I came to respect the power of this blatantly intellectualist approach to music making, I still could find no way to consolidate such a methodology with my personal yearning to be actually *playing* music. No matter how much I wanted it to, conceptual music could not move my soul. Once again I decided to drop out of the prevailing scene. This time I headed south to Mexico.

Mexico offered cheap living, warm weather, and carefree music making. After months of traveling, I finally set up house in the town of San Cristobal de las Casas, very near the Guatemala border. It was an area rich in a very old native musical tradition. Within a few weeks I began to study these traditional Indian songs, both on an old guitar and on a local Zinacantecan pottery flute.

And every single time that I hit a certain high note on that flute, the tom turkey who lived in the yard of my next-door neighbor would let out a single resounding gobble. It was positively uncanny. It was as if the turkey had found its own place in each song, and then joined in right on cue. The third or fourth time this occurred, I ventured next door to meet the very musical turkey face to face. There he stood—fat and brown, red skin drooped over his nose, tail spread wide like a fan. When I began to play the song, the turkey first stared, and then dropped his wings right into the dirt. Then he shook his wings vigorously, raising a small cloud of dust. He advanced step by haughty step in my direction—four steps forward, then four steps back. Every so often, the red wattles on his throat suddenly turned a deep blue color; and then, just as quickly, they returned to red again. And every single time I hit that certain high note at the end of the song's third measure, the turkey let out a single, solitary gobble.

Over the next month, I spent about an hour a day playing strange songs and stranger sounds with that turkey. I learned very quickly that the bird was not actually singing with me, but was, rather, responding to the intensity of the notes. Intensity meant a relation between a high pitch and a loud volume. But this relationship between volume and pitch was never constant, and would some days differ quite dramatically from what I called the "trigger note" of the day before. I speculated that the change was due to a blend of weather conditions, and the turkey's composure. When it was hot, the bird gobbled sooner and more often. Neither was the response directly related to musical sounds. One day a truck sans muffler drove up the street, waking me up from a blissful siesta. From next door I heard the turkey go into one of its gobbling tantrums, like a hysterical child unable to stop crying.

Despite the bird's apparent indifference to the source of any sound, it would, nevertheless, allow itself to be carefully programmed into the body of a particular song. All I needed to do was properly accentuate certain key notes by pitch or volume: *ta ta ta ta TA (gobblegobblegobble) ta ta ta.* And there was method to this madness. If I accented too many notes in quick succession, hoping for a crescendo of gobbles, the turkey soon reached his breaking point, and trotted off in either fright or disgust, as quickly as his two plump legs could carry him. The first time this occurred, a fat woman, with small child in tow, rushed out of her house to scold me in quicksilver Spanish for upsetting her pet. After all, she was fattening the bird for an upcoming Easter dinner, and could not stand by while my frenetic style caused her turkey to lose weight. For my part, it was a rude awakening to learn that my playing companion would soon be served up in the traditional sauce of chocolate and chilies.

Upon further questioning, the woman confessed to me that turkeys like to be serenaded the same way that cows do. "Ride the turkey energy," she advised. "Ride the energy the same way a surfer rides a wave." With that bit of poetic information, she gathered up her dirty-faced little son, and waddled back to her house. But if to the uninitiated her suggestion seems overly esoteric, I myself had a vague idea what she meant. This business was not only about dropping the correct pitch here, the proper volume there. It was also about getting down into the dirt and, if not becoming a turkey, at least looking that bird right in the eye. Granted, if this had been a dolphin, a humpback whale, or a wolf, no one would have any trouble understanding this change in attitude. But this was a gobbling tom turkey, and the entire process had a slightly ludicrous ring to it. In a way, becoming "like a turkey" was every bit as challenging as becoming "like" any of the other, more celebrated animal communicators. At that moment I ceased to *experiment on* the turkey, and instead, began to *play with* it.

Now I rarely brought out the clay flute without first checking to see if the turkey was in the yard. I noticed that his attitude toward me had become much more active. He spent much of his yard time browsing right up against the barbed wire fence, right up next to where I had laid a rug to sit on while I played. One day I invited another human musician to drop by and play with the turkey and me. I taught her a simple, made-up canon, a round on the order of "Row, row, row your boat." Just at that point where the first part ends, signaling the second part to enter, I accented the key transition note with a slightly louder volume. Of course, at that precise moment, the turkey gobbled. The gobble itself added a third harmony to the two human parts within the developing canon. In other words, the three of us were singing a canon with a harmony that would have done justice to Bach. The three of us sat at eye level to one another, singing for at least ten more minutes.

Although this human/turkey relationship was certainly interspecies *music,* I cannot so easily verify that it was also interspecies *communication.* Music involves a sharing of tones, harmonies, and rhythms during a set duration of time. It is a form, and can just as easily be created by a programmed tone generator or a bullfrog as by Bach. The quality, the so-called beauty of the form, of course, lies in the ears of the beholder. Communication, on the other hand, involves much more than aesthetics. It is a transmission—a giving and receiving of understood messages. Interspecies communication is, of course, a subject of much controversy, and inevitably reduces down to the essential question of whether or not an animal, any animal, possesses the consciousness and intellect to share information

with us humans in a dialogue form. Certainly the experiments in sign language with the gorilla Koko have shown that an ape can, indeed, share relevant information with a human being. Similarly, experiments in signing with dolphins in both Hawaii and San Francisco also lead one to the inescapable conclusion that some animals do possess the ability to learn at least the rudiments of human communication.

But all these formalized experiments in interspecies communication share one serious flaw. Every one of them starts by asking the question: Can an animal be taught to communicate with a human being? The animal is the subject, held in a captive situation, and then carefully programmed to learn to give and receive information "the way that humans do it." As such, they fail to take into account that animals may already do it on their own. Thus, we may certainly discover the process by which a chimpanzee learns a humanly derived process leading to the statement: "Me want drink." Yet, once the chimpanzee has learned that obvious request, all we have accomplished is to learn that a chimpanzee, or a dolphin, or whatever animal, can act just a little more like a human being than we had previously believed possible. But if it is truly to be considered communication, then it should also be based upon mutual respect. It must develop as an open-ended dialogue where *both* participants have the equal power to direct the course and subject matter of the learning experience. It is a process that will demand a rethinking of our relationship to the animals. It must include animals as beings of their own environment. In a way, it is more accurately a primitive point of view, and also holistic in its ethics. And in just that spirit, we may then begin to comprehend just a bit more of the animal's wisdom. But of course, this is an extremely difficult proposition given the state of our present human relations to the animals. First, we must either free all our captive animal subjects, or at least give them enough room to roam around at will.

Was I communicating to my friend, the plump tom turkey? To be perfectly honest, part of my mind was quite skeptical about the matter. After all, might not a series of carefully orchestrated truck backfires elicit the exact same response? Thus, a turkey gobble is the stereotyped response to an acoustic stimulus. And perhaps this response is part and parcel of a more general alarm system that has evolved to protect the brood from a sudden, noisy intrusion by some predator. Or perhaps it is also a signal that evolved to warn other turkeys not to intrude on a specific piece of territory. Maybe it is a sexual call, carrying information about the size and fortitude of that particular bird. That would explain why it is only the toms that gobble. It also explains why the bird fled in terror before the very noisy truck. No one, including myself, would propose that a turkey is a prime candidate for advanced research in communication between species. The turkey often seems more the dodo than the Ph.D. There is a famous barnyard yarn about a flock of turkeys all opening their mouths in a rainstorm, and so, drowning.

But there is another level at play here that I cannot so logically explain away. Because, you see, I *did* learn the turkey energy. It is very difficult to say that it is precisely this, or exactly that. It involves sentiments quite peculiar both to our language and to our time. It makes me feel more like a shaman than a scientist; although that admission worries me because of its volatile and easily dismissed characterization. It is about my feelings of collaboration with the bird; about the bird finally sitting beside the barbed wire, waiting for me to appear; my growing sensitivity to the bird's moods—shared feelings about the weather; a dislike of quick movements, sounds, change. It is about learning to operate on turkey time, the turkey dreamtime. And most essentially, it was never the same as studying the turkey's behavior. To the contrary, this relationship was not about *observation,* but rather, about *participation.* It is a difference between all that is static, and all that is dynamic. The essence of nature itself. And if I have failed to explain it as it is, then it is a failing of words themselves. As Alan Watts has written:

> The game of Western philosophy and science is to trap the universe in the network of words and numbers, so that there is always the temptation to confuse the rules, or laws of grammar and mathematics, with the actual operations of nature . . . but just as trees do not represent rocks—our thoughts, even if intended to do so—do not necessarily represent trees and rocks. Thoughts grow in brains as grass grows in fields.

The señora had encouraged me to participate when she so poetically compared the music-making process with a surfer riding a wave. But there are also some critical differences between the two activities. A surfer cannot influence the dynamics of tomorrow's wave by the way he surfs today. Yet after a month, the turkey sat and waited for me, encouraging a continuation of the collaboration. Neither can a wave relate individually to each surfer who rides upon its crest. In a word, a wave cannot relate. Yet the turkey and I eventually formed a genuine bond. And it was a bond not based on food, as might be expected. If this turkey had learned to sign, he would not have said, "Me want drink." Rather, we shared harmonies, rhythms, and emotions—music. I say in utter naiveté: we had become friends.

There is far more at stake in this description of an interspecies relationship than merely personal sentiments of one person's experience with one animal. When I learned to invent music at eye level with a gobbling tom turkey, I seem to have stumbled upon a basic reality of the natural world. This I now call *natural wisdom.* In a way, it is an amorphous concept; at once implying a mystical resonance that occurs between all sentient beings, and yet also encompassing such attributes as responsibility and respect, communication and communion. And until very recently it was acknowledged only as a wisdom of the shamans, and of traditional cultures that had learned to live in harmony with the currents of nature. People who were often considered uncivilized.

And similarly, it seemed as if the prevailing mood in the natural sciences had somehow programmed these sentiments right out of the rest of us. When I, as a layman, first encountered the strange relationship with the turkey, there was simply no outlet for me to express these revelations. Now, on the heels of that other, worldwide revelation of environmental abuse, the tide has begun to turn. We have all learned to take ecology, defined as the study of the interrelationships of living things, quite seriously.

This interrelationship, this natural wisdom, is certainly the major current drifting through the pages of this book. But I am also, upon occasion, far more radical than just a promoter of ecology. I also express very serious doubts about the way that we, as a culture, exclusively license science, with its systematic and logically structured approach, to define nature for the rest of us. The natural world is too important and oftentimes too *unsystematic* a subject to remain in the exclusive hands of our scientists. We must also honor and assimilate some of the so-called primitive earth religions. Similarly, we must offer support to new visions of interspecies community—where humans and animals live in close and equal partnership. We must allow our musicians and artists the same access to nature that we now allow our scientists. They will help us evolve new relationships. Each of these subjects is explored in the chapters that follow. As such we have now embarked upon a journey that will hopefully lead us to a new vision of the natural world.

And in just that style, I left Mexico, and journeyed back to the San Francisco Bay area. There, I convinced a benign radio station, KPFA, to commission me to produce a piece of recorded music with turkeys. Sensing the potential humor of such a strange musical offering, they readily agreed to provide for all my recording needs. On the day of the recording session, held at the Willy Bird Turkey Farm, I experienced the incredible phenomenon of three hundred turkeys all answering a trigger note in perfect unison. I sang the traditional folk song "Froggy Went A-Courtin'"; and every time I enunciated, "Uh huh, uh huh," all three hundred toms responded with a veritable ocean of gobbles.

> Froggy went a-courtin' he did ride Uh huh uh huh (3 times)
> Sword and pistol by his side Uh huh uh huh.
>
> He rode to Missy Mouse's door Uh huh uh huh
> Where he'd been so many times before Uh huh uh huh
>
> Froggy got down on his knees Uh huh uh huh
> He said, "Missy Mouse will you marry me?" Uh huh uh huh.

The edited two-hour recording became "Music to Eat Thanksgiving Dinner By," played over the airwaves at 3 P.M. on Thanksgiving afternoon. It became a kind of new wave Muzak for families gathered together to share the traditional American turkey dinner.

And the music was a hit. Friends encouraged me to continue to explore the connections between music and animals. So over the next full year, I spent time sitting with bobwhites in Ohio, kangaroo rats in Death Valley, and a pack of wolves at a refuge in Nevada. Each species related or reacted to the music sessions in totally unique ways. I was pleasantly astonished when the wolves would cease howling whenever my musical answers to their singing were the least bit off-key. Next, an art patron invited me to the Big Island of Hawaii to try out my emerging techniques with the wild dolphins who lived just off the coast. One day, the dolphins came to me as I played to them. And so, a relationship with a particular species was born. In the years since then, I have developed this musical communication with several other species of dolphins and whales, as well as a score of land animals.

I first conceived of this book as a positive, life-affirming document. One that would demonstrate that all the animals possess a deep wisdom; that human beings are animals too—no better no worse, only uniquely different.

Then I immersed myself in the subject matter, and found myself writing a manuscript that projected foreboding gloom, environmental disaster, and human ignorance. As can already be seen, I came to distrust the working methods of science. After all, to prove my point about animal wisdom, I also had to demonstrate that the scientists had always been misleading us by perpetuating the delusion of the "dumb" animal. But I got seriously mired in my message, no longer able to believe in any of the traditional suppositions about consciousness, behavior, communication, or even that very slippery concept that we call truth. I had kicked the bottom out of the barrel of my source material, yet still actively sought to write a book that would give my ideas a scientific credibility. I wanted scientists to read this book. Worse still, when I waxed positive about the animals as I had grown to know them, the result, by contrast, often sounded like a poorly researched, overly opinionated excursion into cosmic consciousness. I had written a Manichean tract, an alarmist's book—the kind that I never cared to read myself.

So back to the typewriter for a total rewrite. This time I focused the mood of the book more solidly around my uniquely personal experiences with the animals. I invented a fantasy reader, and then wrote the text as a long personal correspondence. My reader was already an animal lover. Her mind was open to the general idea of interspecies communication. She *felt* that animals weren't dumb, but she hadn't really ever thought much about what they *were,* instead. My reader was a practicing zoologist. Both of us loved all those thoughtful TV specials about animals, where the camera looks the creature right in the eye and the animal looks right back. Finally, I wrote the book in the hopes that when I finished it, I would understand my intuitive sentiments just a little bit more clearly. There is hope. And there is also a sense of humor that we learn when we spend time in the presence of turkeys and dolphins.

# 2

# *Touching the Shaman's Myth*

## I. Ducks and Dolphins

A TOTEM IS A KIND of nonhuman ancestor to a tribe of people. The history of a totem-based tribe may include an occasion far in the distant past when the human founder coupled with an animal. The progeny of this mythical union later formalized the cultural aspects of the tribal structure including specific attributes of that founding animal. Today, the modern tribe still remembers the event of their tribe's creation, and so, continues to pay homage to the animal. As the term is used here, *myth* refers to a traditional story of unknown authorship, ostensibly with a historical base, that usually serves to explain some phenomenon of nature, the origins of mankind, or the customs and institutions of a culture. A myth often involves the exploits of a god or a hero.

Certainly not all totemic relations are sexually based. The totem might be an animal who has formed a political or even economic alliance with the human beings. Thus, it is often essential that the tribe and its members both honor and protect the totem animal. In return the animal may teach the human how to live correctly in that particular region that they both call home. The tribe fully recognizes that its animal totem, and by association all the animals in the region, possess survival skills that can well serve a human society. Animals need few if any tools, wear no clothing, utilize locally available foods, and meld harmoniously with the natural rhythms. It was the astute tribe who learned to emulate the ways of an animal in learning their own survival skills. The totem relationship may be thought of as the original environmental philosophy—and perhaps the most successful. It legitimized the relationship with nature as religious, as cultural, and as political. Nature became the first human university. The totem animal became provost.

A totem is often an animal. But it can just as easily refer to a plant, a stone, a wind, a place—even a disembodied spirit. As such, who can ever be certain what prompted Abraham, back at the birth of the Judeo-Christian revelation, to start to pray to an inanimate God who resided in the place called Heaven? And just as we today often pray to a God "the Father" or God "the Son," so the traditional tribe related to their Totem as a kind of older sibling, or perhaps a wise aunt or uncle. Perhaps an example would describe the totem relationship best.

Within the barren interior of Kamchatka, a peninsula of northeastern Siberia, lives a people who are closely related by culture and race to the North American

Eskimo people. The day-to-day diet of these Kamchatka people is generally quite austere. Starvation, especially in the dead of winter, is evidently an accepted and acceptable fact of life. But two times a year, when the huge flocks of migrating ducks fly through their territory, a period of great feasting and celebration begins. Through the centuries a body of music has developed for this ceremonial period, a music derived in both name and melody from the calls of these Aangitsch ducks. The Aangitsch duck is the totem of the tribe. The ducks sing something like this:

It must be noted that the same pitch is not constant for all birds of the same flock. Some sing higher, some lower. What a symphony must be produced from an entire flock of these ducks.

The Aangitsch songs serve two purposes. First, they are essential to the success or failure of the hunt. If the songs are chanted properly, and in the spirit of genuine communion, the ducks lose all fear of men and venture close enough so that a goodly number of them can be killed. This ensures the success of a proper feast. The performance of these songs is a highly ritualized and exacting art, one that requires both a refined vocal technique and a thorough command of what is known as the "duck energy." To the perceptions of the master duck seducer, it must appear as if the ducks themselves demonstrate a kind of altruistic willingness to sacrifice the bodies of some of their avian tribe so that their human cousins might live for another year.

Second, after the hunt has been accomplished, these same Aangitsch songs are sung throughout the period of feasting and celebration, now by the entire tribe. This is best described as a joyous yet serious business. For if the spirits of the slain ducks are not thoroughly appeased in their own language, the flock may not reappear during their next migration. It seems significant to add here that the Kamchatka, like most traditional peoples, never kill more ducks than they need in order to survive. This is no sport. Rather, it is Totemism, a traditional blend of ecology and religion. According to the Kamchatka, the people and the ducks are one and the same spirit.

It may at first seem somewhat awry to realize that the Kamchatka prey upon their own gods, their own "cousins." But on the one hand, it seems identical to the spirit implicit in the Eucharist of the Catholic Church. And on the other hand, it

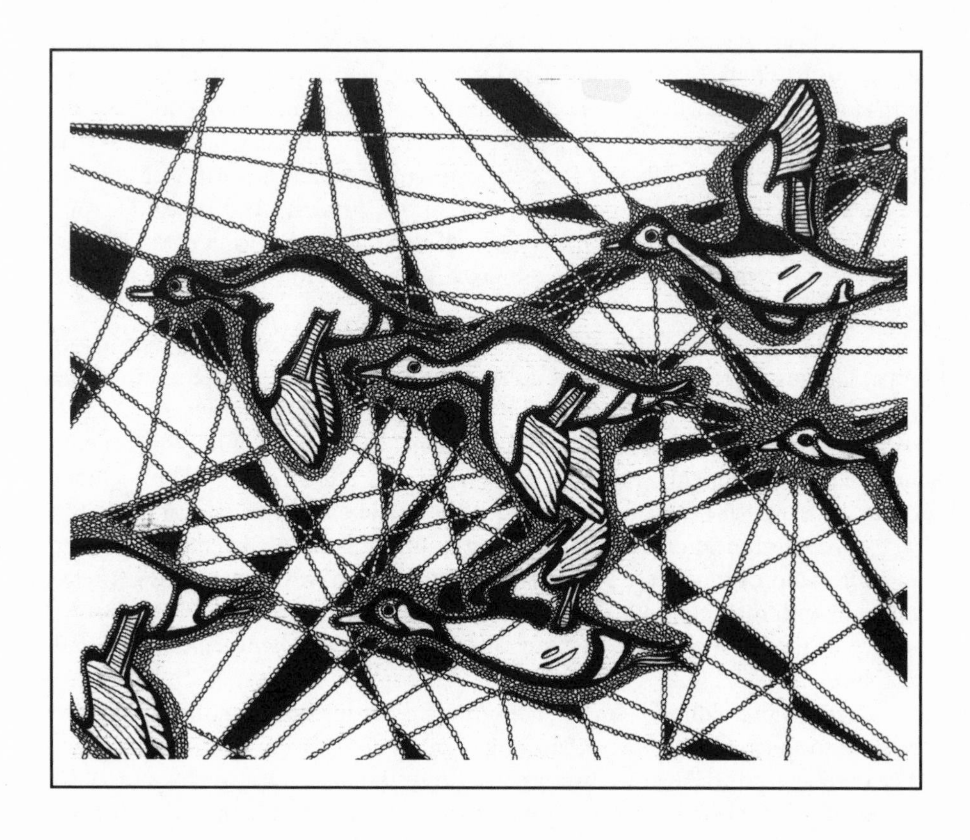

must be noted that the concept of death resides differently within the hearts of these people than it does within our society. During the winter months, when food gets scarce, it is a common practice for the eldest members of the tribe to leave the comfort of the group to walk out into the night to a certain death. This sacrifice, like that of the ducks, assures the continuity of the tribe. And finally, it must further be stressed that a totem is not exactly a god, as we have come to use the word. I prefer the term "cousin protector."

I have heard it stated that there is a modern American "tribe" and that the bald eagle is our totem. But unfortunately for the eagle, and probably the American citizenry as well: State is not tribe, logo is not a totem, a symbol does not a cousin make. Granted, we may too often display the ferocity of the eagle as we believe it to be, but our own ferocity too often exploits nature rather than expressing a balance, a give and take, and a unity with natural process. And of course, our exploitive ferocity has pushed the eagle itself right up to the brink of extinction. We love our creation of an eagle symbology. Yet we never listen to the beat of the eagle's own heart.

The aborigines who live along the tropical northern coast and islands of Australia consider the bottlenose dolphin to be their totem. The elders of the tribe state that the key to understanding the "dolphin energy" is first to learn how the dolphins relate to one another. All relationships between dolphin family members are said to be highly charged with a discrete vibrational energy. This includes telepathy as well as other, more ethereal vibrational bonds. The natural world is a web of communication channels. The dolphin emerges as the networker, the medium who can facilitate a perception of that web. To the aborigines, a relationship with the dolphin is thought to be spiritually enlightening. Here lies one source of the growing myth that surrounds the relations between humans and dolphins.

During late childhood, some of the most sensitive of aborigine youths are taken into spiritual custody by the aging shaman or leader of the dolphin totem. The shaman leads the youths through a rigorous test of physical and psychic ability. Finally, he chooses the most sensitive one of them to become his full-time apprentice. Slowly, over time, the youth is initiated into the complex process of learning to communicate with the dolphins directly. The aborigines of the dolphin totem, or Dreamtime, as it is called locally, view this nomination of the new shaman as a kind of natural analog to the Tibetan system of choosing a tulku. In the latter case, a tulku is said to be the young reincarnation of a recently deceased lama, come back to earth for another round of teaching and leadership. The major difference between the tulku construct, and that of the Dreamtime shaman, is that the latter is often considered to be the reincarnation of a recently deceased dolphin. The shaman is responsible for keeping the line to the dolphins open and active. Thus he is the tribe's representative into the greater web of the natural communication network.

Larry Langly, one such shaman, now an old man, describes his own initiation like this:

> When I was a small boy, I used to go out fishing with my uncle who had studied the dolphins for many years. When he called to them, usually three dolphins would spread out at the sides of the boat and one at the back. They would then make a field of sound and drive a whole mob of fish up onto the shore. He knew so many sounds and the exact way to do it; and sometimes he would take up some salt water in the palm of his hand, and clap his hands together in a certain way to tell the dolphins what kind of fish he was hunting for that day. There are many kinds of dolphins in these waters: the white ones, they live in the sweet waters. In the salt waters you find grey-silver ones and black ones.

Another anecdote is told by Jackson Jacobs. Jackson was probably the only dolphin shaman to be the subject of a television documentary. In it, he relates the fear that he feels—that the ability to communicate directly to dolphins will die with him.

> Every canoe that my Daddy had, he called Dewhn. That means dolphin. One day we were walking along Ggodiggah near Walpardi. My Daddy sang the sacred song even though there was only one dolphin there. I went away, I never believed my Daddy. I didn't take any notice of him singing. I had my own spear and so I went up on the hill and there were all the boys there. And this one boy, he comes over and says, "Hey! You look over there. All them dolphins are everywhere!!" Where they come from I don't know. And at that time we were all hungry because we had no fish. And all of a sudden I couldn't believe my eyes. All the fish, black everywhere in the waves. And my Daddy was singing the song for the dolphins. He walks along the beach and he whistles like this (makes a slow vibrating whistle). We go down there and start spearing them fish, and you know I can see all them fish my Daddy speared. And after that my Daddy does this (slow clapping). And the dolphin comes right up and my Daddy threw that fish to the dolphin.

Jacques Cousteau has documented on film a very similar human/dolphin fish-getting routine, this time among the Imragen tribe who live along the coast of Mauritania in western Africa. The only major difference between the two methods from a half-world apart is that the Imragen, at least in Cousteau's edited version, seem to rely entirely on beating the water with long heavy sticks. The first time that the Imragen allowed Cousteau's team to film the proceedings, the water beatings actually attracted a pod of killer whales. Cousteau surmised that they ventured inshore in hopes of snaring a few of the attracted dolphins for themselves. Likewise, the dolphins stayed far from shore until the killer whales eventually made their exit. Then, finally, a pod of bottlenose dolphins appeared to corral a large school of mullet, driving them toward the beach where the Imragen had strung a net. Next, the Imragen and the dolphins proceeded to "sit down at table" to share a dinner party of the raw and cooked mullet.

Australian aborigine accounts of the same kind of interspecies fishing events often seem to stress the telepathic association between the shaman and the dolphins. The shaman goes into a smiling trance where he speaks to the dolphins, who are often far offshore, directly through the mind-to-mind process. It is the process of the Dreaming.

It can get tricky trying to describe succinctly what Dreaming means to an aborigine. It is nonlinear, rarely a clear case of "this being that." The aborigines assure questioning outsiders that they receive their knowledge about spiritual matters, as well as practical information about how to survive in an otherwise hostile environment, through Dreaming. And likewise, their entire body of sacred fables and creation stories, akin to our Bible, is collectively referred to, all over Australia, as Dreamtime stories. Gary Snyder describes it as:

> a time of creation which is not in the past, but which is here right now. It's the mode of eternally creative nowness, as contrasted with the mode of cause and effect in time, where modern people mainly live, and within which we imagine history, progress, evolution to take place.

My own hunch about the use of the actual term, Dreaming, is based upon the fact that the link between the two realities is so often established through a trancelike state. Probably not a trance, in the sense of modern hypnosis; but rather, kind-of-a-trance or kind-of-a-dream. One such Dreamtime story that relates directly to the human/dolphin connection has come down to us in many varied forms from the Groote Island people of the Gulf of Carpentaria.

The story tells of the "first time" before the present world, when dolphins were smaller and less solid, more dreamlike than they are today. After much internal bickering among various sea creatures, a snail calls in the shark people, who kill most of the dolphins. Then, to pick up the story . . .

> . . . The souls of all the dolphins who were slaughtered became very hard and dry. Finally, after much time passed, they were all reborn on dry land, where they became the first human beings. Never again would their spirits swim quickly through the waves. One night, long after her son had grown, Ganadja (a female) was swimming near the shore when she saw her husband, Dinginjabana, who was now a two-legged man. So thrusting her body high up on the shore, she worked her way over the sand with several heaves of her flippers. And there, Dinginjabana recognized Ganadja as his wife. Ganadja gave a joyful cry and suddenly took the shape of a human being. And today, when we see dolphins strand themselves on the shore, they are almost certainly searching for their human mates.

Over time, the humans Dinginjabana and Ganadja had many children, who became the people of Groote Island. Because these people were the offspring of the wise mother Ganadja, the people of Groote Island are the only human beings who remember that the dolphins are the ancestors of the entire human race. However, *all* the dolphins swimming through the waves today are the offspring of mother Ganadja. And so the dolphins, wherever they might swim in the ocean, have never forgotten that all the two-legged people of the land are their cousins. And that is why, even today, the dolphins seek out their human kin to play as they did during the days of the Dreamtime.

## II. The Gift of the Sticks

In 1978 I became deeply committed to the issue of dolphins that were being systematically slaughtered by fishermen at Iki Island, Japan. The scene seemed straight out of the Dreamtime story—ketchup-colored seas littered with the bodies of dead and dying dolphins. I had been sent to Iki by a half dozen environmental groups in a concerted attempt to put an end to the killings. Not only was I well experienced at orchestrating media events, but also, I had gained a reputation as one who could attract dolphins through music. As these things go, events moved slowly at Iki, but eventually the fishermen came to trust my motives and general good intentions. They told me that if I was able to attract, or frighten, or warn the dolphins away from the severely depleted fishing banks, then they would agree to relinquish their extermination program. The story of Iki Island is quite complex, and will be described in more depth further on.

Just before I left the United States for Japan, I was ceremoniously handed a pair of "dolphin calling sticks" by a friend who represented the Australian Greenpeace Organization. She had been holding on to the sticks for some time, waiting for the proper occasion to either use them herself or pass them on to someone else. She wanted to be very sure that the sticks would be used to aid the cause of protecting dolphins. The Australian aborigines used the sticks for attracting wild dolphins to the shore. In fact, these particular sticks had been constructed by the Oroote Island shaman, Jackson Jacobs.

The two dolphin sticks were eight inches long, pointed at each end, and three inches in diameter around the middle. About an inch from the ends a roughly painted white line circled the sticks. The wood was dark brown, finely grained, and heavy, later identified as Australian ironbark. Testing the sticks in a pool, I was quick to learn that the dense wood sank in water. In general, the sticks were struck together like claves, the Latin percussion instruments used to accent a beat by adding a sharp crack to the rhythm. But, I wondered, are the sticks to be slapped hard against one another, causing a sharp whacking sound? Or should they, instead, be rolled lightly against one another, creating a staccato and very soft gurgle? In fact, I soon discovered three other variations on the otherwise straightforward theme of generating sound with the sticks. I concluded that when and if I was in the water in the presence of dolphins, the correct technique would manifest itself.

Unfortunately, I never got an opportunity to play the sticks at Iki Island. First, working off a large fishing boat never allowed an easy or safe access to the waterline. Second, with well over a thousand other boats working the same area—with all their engines running—it seemed ridiculous to assume a dolphin could hear the sticks over the cumulative roar. Third, and most peculiarly in light of the fishermen's deadly complaint against marauding cetaceans, was the fact that none of us Westerners ever even saw a live dolphin out on the Iki fishing banks. The more I considered the utility of these particular sticks, the more I felt that they had been designed to play from a nice quiet beach, just as Jackson Jacobs himself played them. That was impossible at Iki. Instead, I gravitated toward playing the sticks late at night from the privacy of my hotel room. As with Captain Queeg in *The Caine Mutiny,* the roll of the sticks tended to carry my mind away from the stressful events of the moment.

Six months beyond Iki Island, I finally got my chance to test the sticks from a placid beach on the Big Island of Hawaii. It was the beach of my seminal meeting with the spinner dolphins. I played the sticks for well over an hour that bright, sunny afternoon; sometimes striking them hard together, other times rubbing them or rolling them back and forth as I struggled to keep my head above water. A friend, swimming a hundred yards away, easily registered the sharp thwack while she sat on the bottom in twenty feet of water. At that distance and depth the sticks sounded like someone cracking their knuckles from under a quilt.

The sound was very different from where I hit the sticks together. That same sharp cracking sound seemed to penetrate directly into my body through the solar plexus. It was as if someone were tapping on my abdomen. Similarly, the sound itself seemed to enter my consciousness as much through my plexus as through my ears.

I was very pleased to discover that the cigar shape of the sticks offered a surprisingly low resistance to the water. What on land had seemed an arbitrary form—primitive and, I assumed, determined entirely by aesthetics—in fact demonstrated a well-thought-out hydrodynamic precision. Furthermore, the shape created a resistance that favored the sticks' being bumped up against one another from stem to stern. The resultant sound could not be heard by my friend from more than a distance of twenty feet. But, intriguingly enough, it did sound very much like the well-known "creaky door" call of an echolocating dolphin.

Jackson Jacobs himself may have learned to talk to the dolphins by mimicking their complex echolocation patterns. And while the skeptical, scientific side of me tended to doubt the implications of such a monumentally tricky achievement, the musical side of me wondered if I might be able to learn such a thing myself. After all, how did he coax those dolphins into corralling fish into shore? Reading an interview with him, I inferred a subtle mixture of music and telepathy. And isn't it also true that many telepathic adepts claim that the messages enter and leave through the solar plexus? Could there be a connection between the "creaky door" sound and the fact that the loud thwack had made me so aware of my plexus? Could I also learn to communicate directly to the dolphins, possibly get them to corral fish for me? Without a teacher it might take an entire lifetime to learn. I had a lifetime. For such an endeavor I would need to cultivate determination, patience, and perseverance: positive traits no matter what the exercise. These dolphin sticks were leading me to believe that such a task was no longer the case of a hopeless dreamer trying to pull fantasy relationships out of thin air. Jackson Jacobs himself would have probably demanded that one must be a hopeless dreamer before one can even begin.

Whatever my speculations about human/dolphin communication, that day in the water with the dolphin sticks I was unable to attract any dolphins.

Unfortunately, the aborigine dolphin sticks disappeared from my possession before I got a chance to use them in the vicinity of dolphins. Then, later, I also heard a rumor that Jackson Jacobs was very sick, and may have already died. There was nothing else to do but try to make a set of sticks on my own.

By now I felt somewhat knowledgeable about most of the explicit design elements. Or to be precise, I possessed the ability to carve a pair of wooden sticks that would be able to generate two or three specific kinds of sounds underwater. Was it that simple? Probably not, but then I had nary a clue about Jackson Jacobs's stick-making technique. I decided that here was a classic case of the chicken and the egg. In this case, to make a set of dolphin sticks was to learn how to play them—and vice versa. It was all the same thing.

Then there was the tricky matter that my former sticks could only be construed as genuine "power objects." On the simplest level, this meant that anyone and everyone who ever saw the sticks, from skeptic to believer, felt something very special because these were genuine aborigine dolphin sticks made by an aborigine dolphin stick maker. How could my own ever compare? Had Jackson Jacobs imbued his sticks with a ceremonial magic that was beyond my ken and culture?

Fortunately, it never seemed especially important whether or not I was privy to Jackson Jacobs's brand of magic. He had believed in it. It had probably governed every stroke of his cutting tool, the way he walked down to the beach, the way he sang, the manner in which he smiled back at the dolphins. Certainly, he had breathed some ritual spirit into his task, a spirit that had imbued the original sticks with a power we Westerners call by the name of magic. And certainly, Jackson Jacobs's wooden power objects, struck together by a master, were able to attract and even talk to dolphins. How many Ph.D. biolinguists, all using their power object concrete pools, and their power object computers, have failed to achieve even an atom of Jackson Jacobs's astonishing results? On the one hand, I revered his power, yet on the other, I trusted my own ability to achieve a modicum of success with the resources at my disposal. What purpose would it serve to give up because I was unable to mimic or emulate an unavailable cultural ambience? In a very real sense, my choice was whether or not to attempt the route of the dolphin sticks or the route of the computers. I'd stick with the sticks.

I surmised that the power in the original sticks was symbolic, historical, and highly ritualized. Yet, as I say, it seemed patently naive and even vain to attempt to copy the unknown ritual sentiments of aborigine culture. It seemed somewhat akin to practicing open-heart surgery after watching it done on TV. Instead, I would do much better to search for traces of my own Western, late-twentieth-century belief system. Such symbolic elements would serve to focus the mind of the player to the enormous task of attempting to communicate to another species. But what did I as a musician, as a university-educated white man, truly and deeply believe in?

First, I believed in time. That is, I myself took special care when handling handmade objects that had taken a long time to make. Time equals care; care meant responsibility. And once a person felt that any object was his or her responsibility, then that object had also acquired no small measure of power. So in constructing two dolphin sticks, I knew that I needed to take much more time than our society would consider normal to the task. I would carve these sticks as Michelangelo had painted the ceiling of the Sistine Chapel.

Second, I believed in sincerity. That is, I had to sincerely believe that whatever wood, shape, and symbols I chose to carve on the sticks was done in the spirit of their eventual use. This is also known as a metabelief: the belief in one's beliefs. Metabelief is another word for sincerity. But the very idea of a power object seems very far afield from my culture's belief system. How does a person learn to believe in a system of beliefs that one has been taught from a very early age to disbelieve?

The answer is: third, I believed in intuition. After all the homework has been laid aside, after all the source material has been carefully digested, after anyone and everyone has been given an opportunity to air their opinions about the strange task of building dolphin sticks, then and only then would I let my intuition be my guide. On the one hand, our society tends to disbelieve and then either trivialize or capitalize anything so primitive and esoteric as a power object. After all, a power object simply does not fit into the usual categories that we moderns have invented for the concept of tools. Tom Robbins satirizes this well when he writes:

> A disbelief in magic can force a poor soul into believing in government and business.

Yet on the other hand, our society is full of its own potent power objects. Jimi Hendrix's guitar, Groucho Marx's cigar, the Dalai Lama's robes. If my sticks worked to call dolphins, they would definitely be in that class. And the best way to combat any gnawing self-consciousness implicit in such a task? I simply gave precedence to that great Western cultural context known as art. I would wear the hat of the artist. Furthermore, I learned to accept my intuitive trust in the often remarkable powers of human intuition. Here is one more "meta": this time, meta-intuition. Meta-intuition means nearly the exact same thing as common sense.

In the case of the now formative dolphin sticks, the basic shape was plainly prescribed by the shape of the originals. However, I quickly decided to flatten out the waist, giving to the sticks a wider slapping surface, as well as a better fit to my small hands. Australian ironbark was not to be found anywhere. No matter. After much discussion with a cabinetmaker, I chose an Amazonian hardwood known as purpleheart. It differed from the original wood in two important ways. First, the sound was even more resonant, if not a bit more glassy. Second, the purpleheart, being not so dense, floated. I guessed that Jackson Jacobs chose ironbark because it was the most resonant wood available to him. Since he played his sticks from the beach, he never had a fear of losing them in deep water. I would probably use my own sticks from the platform of a small boat in deep water.

Several months passed. By now, the purpleheart sticks had been roughed into their final shape. At this point I decided to carve some symbols directly into the surface of the wood. I reasoned that this would make one side of the sticks quite bumpy when they were rubbed together. More bumps meant not only the production of more creaky clicking, but also the opportunity to create an actual obbligato clicking, if the sticks were rubbed the same way over and over again. Furthermore, I felt that the appropriate symbols would aid to focus the thought processes of each person who played the sticks, whether they comprehended the hidden meanings or not. In other words, the symbols would provide an aura of meaning beyond the musical. First, they would set a mood of mystery for any single stick-playing occasion. Second, they would serve as reference points for an entire world of further study.

A picture of a dolphin seemed the obvious symbol for one stick. But what kind of dolphin? An anatomically perfect dolphin was out of the question. Such a rendition might permit an unimaginative person to identify a "real" dolphin; but then, the real dolphins were meant to be swimming in the real water. Neither did a stylized aborigine dolphin serve my purpose. If anything, their art seemed too culturally specific. Yet taken out of that cultural context, the aboriginal dolphin looked too much like a duck. I wanted my dolphin to be recognized at a glance as a dolphin—but not as a copy of some other culture's picture of a dolphin.

Finally, after much late-night sketching and books full of glossy photographs of primitive art strewn all over my dining room table, I hit upon a dolphin that was unique, if not vaguely derivative of Mayan and Haida styles. The head of the dolphin contained three beanlike projections representing the brain, the melon (a fatty lens located in the forehead that helps to transmit sound), and the lower jaw (similarly employed for receiving sound). As a final touch to the finished drawing, I added a fetus growing within the mother dolphin's womb. That was enough for one stick.

The other stick posed a quite different problem. At first, I wondered if the obvious drawing might not be a similarly stylized human being. Hit the sticks together, and you symbolically join human and dolphin to produce the sound. But somehow, the human motif did not seem correct. A human being was already hitting the two sticks together. Another human bespeaks a preponderance of humans. Overpopulation, people hitting people, all making entirely too much noise.

I immersed myself in the task, poring over books of hieroglyphs, signets, and runes, often feeling like Mickey Mouse as the Sorcerer's Apprentice—delving into matters far beyond my meager experience. But slowly, over several weeks' time, first my eyes and then my intuition began to settle upon various historical symbols representing interconnectedness. First I drew five interconnected circles. No, that was the Olympic Games. Each day I subtracted one circle and stared at it for fifteen minutes. My favorite was the simplest: two interconnected circles.

Weeks later, I found to my great satisfaction that the two circles had once been used by hobos during the period of the Great Depression. Evidently, hobo subculture had developed an entire lexicon of signs that were drawn on sidewalks and posts as a way to signal other hobos. The symbol of the two interpenetrating circles meant: "Don't Give Up." The next day I incised the message onto one of the dolphin sticks.

For a second symbol, I chose the Mayan pictograph signifying the fifth direction. This direction is, in fact, the center—the center of the universe, the center of any object upon which it is incised. But it is more than that. By writing this sign upon an object, one focuses the entire universe upon that point, a Native American way of writing, "Be here, now." The sign itself represents a snail whose head points back into its own shell. I carved the snail just below the hobo sign.

But whenever I read the two symbols together, they always seemed to scold me: "Take your time, and do your best, and the dolphins may come." It sounded a bit too much like the directions to a college entrance exam, the punch line to one of Aesop's fables. If, like Jackson Jacobs, I had succeeded in infusing the sticks with some invisible spirit, then he seemed much too much the effete moralizer. He lacked a sense of the great unknown. I longed for a message that would suggest the human/dolphin bond as the essence of a profound mystery. And it was just at that moment that I stumbled upon the religion of the Dogon of Mali.

The Dogon believe that human civilization was carried to earth by the god Nommo, who flew here from a small binary companion star of Sirius, in Canis Major. This tiny star is known today as Sirius B. Before Nommo's arrival on earth, all of humanity was steeped in ignorance. Nommo taught us language, art, agriculture, mathematics, and writing. Significantly, the Dogon describe Nommo as an ocean-living, air-breathing creature, possessed of opposing flukes just like a dolphin. This fact is especially noteworthy because the Dogon live in the middle of the Sahara, over a thousand miles from the sea. But you draw them a picture of a dolphin and they say Nommo.

If at first, this description of the Nommo myth seems similar to a hundred other traditional stories, then it possesses one more element that forces all the astronomers among us to sit up straight in their chairs and scratch their heads in befuddlement. You see, Sirius B, the companion star of Sirius, is totally invisible to the human eye. It had not been discovered by astronomers until the relatively recent invention of powerful telescopes during the early years of this century. There is no conceivable way that the Dogon, or anyone else for that matter, could see Sirius B.

The upshot is that the Dogon have never declared that they *see* Sirius B. Rather, Nommo told them that the star was his homeland. He even went so far as to draw them a diagram that showed how the twin Sirius stars revolve around one another in a fifty-year cycle. Remarkably, when Nommo's drawing of the orbit is placed alongside a recent computer-enhanced rendering of the same orbit, the two appear identical.

But of course, such an astonishing claim of extraterrestrial visitation is bound to encourage detractors. Carl Sagan points out that the two French anthropologists who first recorded the myths and religion of the Dogon in the 1920s had been preceded in Dogon country by other Europeans, fifteen years earlier. One of these earlier explorers had been known as somewhat of an amateur astronomer. And at about that same time, the recent discovery of Sirius B had been causing quite a stir within astronomical circles owing to the fact that it was the first binary star yet discovered. Thus, concludes Sagan, the Dogon worship of Sirius B is nothing more than recent astronomical front-page news transcribed into mythic terms by a primitive tribe during a fifteen-year interim.

DOGON

I find Sagan's conclusions less than adequate. He seems a bit stuck in a reductionist approach that starts from the premise that, indeed, this Nommo business cannot be credible, so let us, instead, settle upon a more rational explanation. But I wonder if Nommo can be so cavalierly debunked. There are some researchers who believe that Dogon cosmology not only precedes the earlier amateur astronomer, but that it goes all the way back to the ancient Egyptian worship of Isis. Isis, the principal female deity of Egypt, was goddess of Sirius. Osiris, her husband, is sometimes depicted as her "dark companion."

But it is not my intention here to debunk the debunkers. Today, we live out our lives within a culture compulsively prone to analyze, draw conclusions, and pass judgments. Consequently, many of us have lost what is in truth a very healthy regard for the power of mystery. Puzzles exist only to be solved. It therefore seems appropriate sometimes to take a step backward and simply applaud the wondrous inscrutability of it all.

I did not get a chance to discuss the matter of dolphin sticks with Jackson Jacobs before he passed from this life, but I've heard from a few sources that he himself believed that the sticks were merely a vehicle for unleashing a telepathic connection that exists between sensitive human beings and all dolphins. Unfortunately for me, Jackson Jacobs was a direct heir to a telepathic tradition that dates back at least several thousand years. My upbringing did not involve dolphins at all. Telepathy was never mentioned. But I have sensed some inexplicable mind-to-mind connection while playing music with animals. I do not know yet how to channel my fleeting sensations.

By comparison, Jackson Jacobs seemed almost blasé about his ability to engage the dolphins by hitting two sticks together. The dolphins respond to this rhythm by corralling fish for the humans. I envision great swarms of malnourished aboriginal children wading into the shallows to receive their only meal of the day. In my case, I feel no need to ask the dolphins for some fishy handout, although I am the first to acknowledge the dinner table as a powerful social lubricant in any tongue. Unfortunately, Jackson Jacobs has chosen to leave the scene without providing any hint of his musical technique. His unbearable silence has taught me to venture out onto the waters by myself. Build some sticks, find some dolphins, and get to it. The hobo sign reminds me not to give up. The Mayan pictograph scolds me whenever my mind wanders, as it is so wont to do.

This morning, after six weeks of meditative carving, I have completed my task of inscribing the Dogon map of the fifty-year cycle of the binary system of Sirius upon one of the dolphin sticks. I carved it just beneath the pictograph. And last night, I stepped out into the cool night air to search the skies. Sirius is the brightest star in the heavens, and not difficult to locate. Mostly, I thought about the constellation that surrounds the star: Canis Major, the great dog. I saw no great dog;

in fact, when I connected the dots, I was left only with a great lump. Every so often Osiris/Nommo flitted through my mind. Sirius B was nowhere in evidence.

And this afternoon, I unveiled the sticks to a friend who dropped by to visit. "What do all these squiggles stand for?" he asked good-naturedly. The idea of symbols "standing for" something, like a patriot saluting the flag, gave me a brief moment's pause. "Oh, I don't know anything about that," I answered coolly, thus declining a long-winded explanation about Mayans and hobos and Dogon. "Why don't you just play them for a while, and tell me what you feel."

He stared askance at the symbols for an inordinate amount of time, running his thumb over and through the deeply cut incisions. Then he shifted the sticks in his hands, at one point throwing one of them into the air to better savor its strange heaviness. Had the wood been dyed? No, I answered, deep violet is the actual color of purpleheart. He stared a moment longer, and then very gently bumped the sticks together. "Click." I was pleasantly surprised; for some reason I had expected him to slam them. Then he rolled the sticks back and forth along their carved surfaces. The sound possessed much potential as an approximation of the "creaky door" sound.

He placed the sticks down on my worktable, taking special care to ensure that they would rest just parallel to one another. He peered at me with the faintest hint of a grin, stared at the sticks again, and looked at me again. Then he moved very quickly away from the table. No comment. Silence.

Two months later I found myself on the Mexican Riviera, directing a well-funded dolphin communication project. I had been hired to attempt a long-term relationship with the free-swimming spotted dolphins who frequent that azure seacoast. The very first day of the project, I gathered up my dolphin sticks and kayaked out into the large autumn swells. Within an hour, the dolphin sticks had been washed overboard while I worked to keep control of the unstable boat. I never found the sticks.

Anticlimactic? Of course. Discouraged? Not really. Perhaps the spirit of Jackson Jacobs had been trying to tell me that I had overlooked some unknown but crucial detail. At the time, I knew that I would be spending six precious months working to develop a relationship with dolphins. Maybe that would give me access to the detail. I could always make another pair of sticks in a year or two. I had never much liked that garish purple color anyway.

# 3

# *Kivvers and Iki Island*

AT NINE YEARS OF AGE I had become an enthusiastic fisherman. Nearly every summer's day found me up to my knees in Lake Cochituate, casting brightly shiny lures into the depths for hours on end. Once or twice that summer I hooked and landed my first trout—a ten- or twelve-inch rainbow. They were treated as rare treasures, proudly held up to anyone who cared enough to look. After trout, yellow perch were accorded second stature. I judged them slightly more desirable than white perch, although the latter fish often grew much bigger. Bass? Pickerel? At nine years of age, these fish were still two years away from being anything more than exotic pictures in a fishing magazine.

Inevitably, I caught sunfish. There were three species. The first, the "punkin-seed," looked nearly tropical in its bright orange, blue, and green skin. Second were the "crappies," sometimes as much as twelve inches long, and very brown, very camouflaged within their natural habitat. In all of nature, the only animal to be handed a more insulting name than the "crappy" is probably the killer whale. "Crappies" were absolutely delicious when fried up in a batter of cornmeal. The third variety was nearly square in shape, with a dark blue flap covering its gills. The identification books called them bluegills. I called them kivvers. They were the bread-and-butter fish for a nine-year-old seeking to take a prize home to mother.

Kivvers would bite at just about anything dropped into the water. I remember how my friend Donny once snagged three or four with a small safety pin jiggled back and forth in front of their faces: The two of us would stand on his dock and stare into the water. Nothing in sight. Then we'd bait up, throw the line over the edge of the dock, and watch with glee as the kivvers materialized around the bait. They'd zoom in for the kill like piranhas attacking cattle in the Amazon. At nine, I found great satisfaction in bringing home the two or three biggest kivvers of the day. The beautiful little fish rarely tipped the scales at eight inches and twelve ounces. I'd sling them over the handlebars of my bike, and ride home with them flopping in the breeze. Finally back home, I'd rush in and hold them proudly before my mother, who'd always receive this gift from the deep with unerring enthusiasm. She'd place them on the kitchen counter for a few minutes, only to return to the kitchen to pick them up delicately by the tail and deposit them unceremoniously into the garbage pail. At the time I entertained no thoughts of throwing the fish back into the water. I simply had no sense, whatsoever, that the death of a kivver was a life taken in vain. And yet, for some strange reason, neither my mother nor I ever considered eating the kivvers, although they were close relatives of the very delicious large-mouth bass. During the summer of my ninth year, our garbage can must have received in excess of fifty kivvers. Another name for kivvers is "trash fish."

Two years later, at eleven, I had turned into an old jaded professional of a fisherman. By now I had graduated to casting oversized floating plugs for bass, and exquisitely tied flies for trout. My attitude had also changed quite markedly. Now, the bass and trout that I caught with regularity were more an issue of a tasty dinner than of any kind of reward for mother. And naturally, kivvers, who continued to bite at anything, no matter if it was a plug bigger than itself, had become a nuisance. All the sunfish were bait stealers. As far as we were concerned, they were no better than horse thieves in the old West.

But once in a while, fishing from Donny's dock, we would spend hours fishing without any luck. Finally, out of frustration, we'd both switch to worms, and begin one of our kivver-catching contests. It was fast and furious work. Sometimes we'd see who could catch the most in a minute, or five minutes, or a half hour. One of us would hook a fish and slash the fishing rod directly backward with all our might, whipping the fish out of the water and over our heads. The object was to use enough force so that the unfortunate animal would rip its mouth away from the hook, without swallowing the bait. The kivver would land twenty feet back in the bushes, as the bait was immediately cast back into the water.

I recall one particularly gruesome afternoon. The two of us were more bored than usual. It was a hot steamy Fourth of July weekend. We had already conceded the day to the trout hours ago, and so, switched over to hooks and worms. Fifty or sixty kivvers were flopping around in the bushes like so many Mexican jumping beans. Likewise, the local cats were out in force, swatting the kivvers back and forth, pouncing down hard every time a fish jerked or thrashed.

Donny produced a bag full of firecrackers. We called them cherry bombs. They were not the kind that goes crackity-crack down a long fuse. No, cherry bombs were genuinely dangerous, miniature explosives masquerading as toys. I picked up the largest, liveliest kivver I could find. It was a magnificent creature, an orange sunburst belly abruptly clashing with its aquamarine throat and its blue-black gill cover. Donny pushed a cherry bomb down to the back of the kivver's throat. I lit the fuse, which dangled out of the fish's mouth like a cigarette. He dropped the fish back into the lake. There, it floated on the surface for the briefest moment, as if it were trying to get a bearing on the situation. Then it darted away and down, leaving a trail of bubbles from the burning waterproof fuse. No, there was no explosion. The fish was already too deep. Instead, the glassy surface of the lake suddenly began to shimmer, just over the spot where the fish had disappeared. It was as if some vast, forgotten undersea monster had been rudely awakened.

This monster lay mercifully hidden from consciousness for many years, only to be dredged up before my eyes during a tense, bloody weekend at Iki Island, Japan. For an entire afternoon my crew and I sat in a boat observing fishermen go through the systematic paces of their dolphin extermination program. We were a group of American environmentalists sent over to end the slaughter; now we were being given a graphic demonstration of just what we were up against.

On one side of tiny Tatsunoshima Bay, helpless dolphins drowned in the slack pockets of carefully strung fishing nets, in a futile effort to escape. Fishermen waded into the ketchup-colored sea to lasso the animals one by one. The rope would be transferred to another group; ten or more men heaved to the ropes, slowly but persistently dragging each dolphin up onto the fine white sand. There, another group of fishermen stabbed the dolphins repeatedly in the throat with long-handled spears. The dolphins would begin to thrash as their lifeblood spurted in rivers down the beach and back into the bay again. Then as the flow began to ebb, the thrashing would metamorphose into a stiffly tense shimmer. A moment longer and the body would relax again, now quiet in death. Then, still another group of men would drag the carcass along the beach, where it was tied with ten or more carcasses to another fishing boat. When the quota of twenty dead dolphins was reached, the boat motored across the bay to the fishing building. There stood an incredible machine going through its paces just to the rear of the shiny modern building. The dead dolphins were deposited on a conveyor belt, dropped into the maw of the beast, and finally emerged as fertilizer and pig feed.

The monster I saw that afternoon was a reflection of myself in the face of the exhausted Iki fishermen. The exploding of kivvers and the stabbing of dolphins seem equal in their unmitigated cruelty. On another level entirely, both deeds were perpetrated upon "trash fish," who were guilty of impeding human beings from catching the more highly prized food fish. The Japanese called their dolphins "gangsters." We called our kivvers "bait-stealers." Looking into the eyes of a fisherman as he ripped the net off a dolphin's back, nearly severing the dorsal fin in the process, it was the same look that I saw on the face of eleven-year-old Donny. It was the same look that I, no doubt, possessed as well. The eyes, it was that same look in the eyes: intense but also futile; glazed, like the eyes of an already dead animal. That dead stare ripped me away from the present trauma of beloved dolphins, and back to the hidden memory of an eight-inch kivver diving with a lit fuse dangling from its mouth. That glazed look ruthlessly cut through all the differences of culture, politics, and economics upon which this Iki issue was so solidly based. I stood there, transfixed upon the bow of an Iki fishing boat, silently observing the slaughter of a hundred writhing dolphins, recognizing that there was an otherwise innocent eleven-year-old standing there, partaking of the same nightmare.

That evening I confessed the story of the kivvers to two of my Japanese colleagues, as we sat on the tatami-matted floor of a hotel room, sipping hot sake to ease the pain of our experience. Atsumi, who served as the group's interpreter, answered my confession by telling us how he used to pull the legs off beetles at about the same age of eleven. Sometimes he'd improvise by nipping off half a wing, watching the insect whir around in circles, unable to steer a straight course. Tetsuo, a newspaperman stationed on the island to report on our progress to end the killings, drew on his own well of stories. He told about collecting swarms of tent caterpillars, and then incinerating them alive over an open fire. He further confessed to a special relish he felt when one of the hapless creatures sizzled and finally popped. The three of us sat there in the half-light of dusk, contemplating through the window of our sake glasses the universality of human brutality to animals. We concluded that nearly everyone must possess similar gems from the hidden treasure chest of childhood memories. Tetsuo added that it was an affliction of little boys more than girls—an expression of male children playing at the competitive power struggle of a patriarchal society. We teach our children that nature is there to be exploited, that the animals exist for our utility. I was moved to find a book I had been reading and thus quote the famous words of Henry Beston:

> We need another and a wiser and perhaps a more mystical concept of animals. Remote from universal nature, and living by complicated artifice, man in civilization surveys the creature through the glass of his knowledge and sees thereby a feather magnified and the whole image in distortion. . . .

The only difference between ourselves and the Iki fishermen was that as we had become adults, we had somehow outgrown the utility of glazed brutality. A commercial fisherman could not so easily retire from his own brutal nature as could a musician, an interpreter, or a journalist.

Atsumi explained that Japanese schoolchildren are taught that *all* animals are to be treated equally. In fact, to kill a rat is no different than killing a cow, no different than killing a dolphin or a hundred-ton blue whale. Only humans are different, somehow exalted and apart from the otherwise unilateral ethic of killing and dying. This is basic Buddhism as taught in the classrooms of a basic Buddhist nation. Atsumi was right. We environmentalists had fast learned that we would never get anywhere in Japan if we tried to isolate and then condemn the Iki fishermen for the specific act of killing dolphins. There could be no censure from a culture that treats its own citizens like a big extended family. We environmentalists were foreigners, outsiders, chastising the hardworking fishermen over the incorrect sentiment that dolphins were somehow "special." But as was so often told to me, the Japanese never even ate meat as anything but a condiment until the Americans started selling beef to them in the middle of the nineteenth century. It was all too confusing. First we promote the eating of certain animals, then we condemn them for eating those they already eat.

Whales and dolphins were never accorded the culinary exempt status of lamb, beef, and pork within the culture, for the simple reason that the Japanese sincerely believed they were eating a fish. A very large fish, even an air-breathing fish, but a fish nonetheless. This identity adheres even today as proved by the fact that the Japanese whaling industry is governed by the Ministry of Fisheries. There is, as of this writing, no Japanese Marine Mammal Commission. But despite the Japanese bureaucracies' ongoing struggle to continue to regulate their cetaceans as fish, despite massive worldwide protest, most of the ordinary citizens in Japan are by now aware that it just is not so. But how did they learn that whales and dolphins are indeed mammals, to be awarded a different ethical status? One Iki fisherman told me that he and the rest of the local people had learned by devotedly watching *Flipper* on television, every Saturday evening.

It was almost too crazy a way for the fishermen to be told that they may have been sinning against an old Buddhist precept. And yet, it somehow made it just a bit easier to disregard the zoological fact amid all the other trappings of an American Hollywood fantasy that identified dolphins as also a marine version of Rin Tin Tin. Flipper was certainly nowhere in evidence among the "gangsters" who competed so vigorously for the exhausted fish supply out on the local bank.

Then there was the patently fascist matter of American environmentalists arriving on Iki while American tuna fishermen annually caught and killed between ten and a hundred times as many dolphins as the Iki people. Thus, each fresh environmental volunteer thrust upon the Iki stage usually found himself face to face with Mr. Obata, the Fishing Union officer in charge of the dolphin roundups. Like the mythical Sphinx, Obata stared inscrutably into the eyes of each of the young Americans, and then jabbed with his inimitable question: "Why have you come to Iki when your own fishermen kill ten times more dolphins?" It was a tough question. First off, very few environmentalists considered the San Diego fishing industry "their own." Second, though Iki was thousands of miles from the USA, it was still much more accessible than the American slaughter, which went on out of sight on the open sea. And just like the Sphinx, Mr. Obata killed any hopes of a future audience should the young man before him answer so typically: "You people kill dolphins on purpose. The Americans kill them by accident." After all, was it an accident that the tuna fishermen had begun killing dolphins only when their greed forced them to change from the benign method of fishing with hook and line, in favor of the larger and less choosy purse seine?

Another answer, which, strangely enough, met with much more favor with Mr. Obata, was: "We love dolphins, and will go wherever they are being killed." This he could understand. The Iki fishermen were ostensibly religious men. We Americans had arrived on Iki Island on a kind of new-age pilgrimage. It was a cult. Those who belonged to the cult were called "dolphin lovers." Not coincidentally, the very concept of love had first slipped into the mainstream of Japanese culture via the proselytizing efforts of sixteenth-century Christian missionaries. To this day, it still retains faint traces of its original foreign accent. When Mr. Obata called an environmentalist a "dolphin lover" his comment carried with it no small measure of poorly concealed derision. It sometimes made me feel as if I were a saffron-robed Hindu admonishing the president of a Chicago meat-packing company to stop killing cattle because: "Please, kind sir, you are very bad indeed to be killing one of our many gods." But because so many environmental crusaders had been arriving on Iki shores for so many years in a row, all preaching the preservation of dolphins, all of us were, by now, branded as dolphin lovers. This, whether or not that single overriding fact had anything at all to do with our purpose on Iki.

Then there is the demographic issue of extinction. It is certainly true that the Japanese continue to espouse their "right" to hunt the great whales, despite the fact that *all* species are endangered. But fortunately, none of the dolphin species of the Iki area are yet on the verge of extinction. Unfortunately, too many environmentalists have argued that the Iki dolphins are nearly extinct. Yet every census ever conducted out on the fishing banks has shown that there are at least thirty thousand dolphins swimming through the area each winter. Of course, that also means that the fishermen have depleted the stocks by at least one third; but that is no argument at all to a fisherman who has a dolphin problem.

Finally, there is the confusing metaphysical concept of extinction itself, another idea laden with much subtle and almost subconscious cultural bias. Japan is a culture in which reincarnation is part and parcel of the process of life and death. We are concerned here with souls, not mere species. Likewise, there can be no ultimate censure in a universe in which Karma accrues, and then eventually exacts its retribution. The possibilities for speculative permutation can become breathtaking if not literally infinite in their scope. For example, perhaps a stabbed Iki dolphin gets reborn as an Iki fisherman. Then when that fisherman dies in turn, his soul is once again reborn as, you guessed it, a dolphin. Only to die again, to be reborn as a . . . mitochondria? Here is a mystical cycle of nature. Ecology and religious belief interpenetrate to become a single reality. After the first great slaughter of dolphins at Tatsunoshima Bay, the fishing union erected a stately memorial to the souls of the dead animals. It said, in effect, "We beg the departing spirits to forgive us for this terrible deed." And likewise, any Westerner possessing a sincere desire to stop this human/dolphin conflict first had to fully recognize this underlying natural philosophy before rudely charging forward, brandishing dogma that asserted the so-called "self-evident facts"—such as the one about the exhaustion of gene pools. Or even that other ugly possibility of a world "without" whales. Otherwise, one fell prey to another worldwide myth—the one about the "ugly American."

Of course, while trying to attain a balance among all these alternative definitions for seemingly universal truths, it became essential for any pragmatic environmentalist to treat the Iki fishermen not only as peers, but even as teachers. Despite the fact that the term "dolphin lover" carried with it overtones of both a nature cult and even racism, we still had to own up to the fact that we Westerners had arrived at Iki Island because, indeed, we were dolphin lovers. That was why we came to Iki, and not to Kenya to save elephants, or Alaska to promote wolves, or in fact, dozens and dozens of places worldwide where animals are being slaughtered to make room for human territorial expansion. If Mr. Obata taught me anything, it was that I had given the dolphins a special power to move, specifically *me*.

Iki Island taught me that there are people all over the world who share this same strong totemic relationship with dolphins. Of course, in the beginning, the world media was attracted to Iki because of the graphic image of innocent and smiling dolphins being ruthlessly led to a vivid bloody death. But the fishermen saw what was happening. They disallowed any further access to the continuing slaughter. But if it went on out of sight, it certainly did not go on out of mind. The journalists showed up anyway. The basic issue of *dolphins* became more than the Iki Islanders had counted on. It was somehow unique—a magnet attracting environmentalists, journalists, photographers, students, and scientists—all of them pilgrims in the cause of dolphins, all of them sociologists trying to get a grasp on what might prompt any human being to kill a dolphin. Later, in Tokyo, I was to hear a representative from Amnesty International protest that there was more worldwide interest in the Iki dolphins than in the Vietnamese boat people, who were having their own troubles at that same time. It may have been true.

Some tried to explain the magnetism of Iki Island as a worldwide condemnation of the killings. The slaughter was too clearly an act of supreme ecological negligence (which it was). Others felt that the solution to the problem of a fishing economy out of kilter always seemed so close that the journalists expected it to be solved at any moment. It was never so easy, but it may have kept a few of the journalists on the scene. But neither of these reasons do any more than equate dolphins and Iki Island with all those other animals in all those other places. Iki was larger. It seems that in this last quarter of the twentieth century, more people love dolphins than just about any other wild animal.

It is not only the tantalizing conjecture that the creatures may possess more intelligence than any other animal, including humans. The issue of intelligence certainly adds to the general mystique, but it is not the heart of the matter. No, it is more than big brains, more even than the distinct possibility of interspecies communication. It involves equally their qualities of warmth, animation, and physical appeal. That irresistible dolphin smile has won us over. I wonder how I might have altered my grim relationship with the kivvers had they possessed such a grin.

And most remarkably of all—these intellectually dynamic beings seem to love human beings. We need not always coerce them with food as we do for dogs and cats and horses. Dolphins are not pets; they seem more like peers. Yet they remain wild animals. Sail a boat out onto the ocean. Eventually, you are sure to spot dolphins riding the bow wave. But, I have always wondered, how can it be true that these very aware creatures are not afraid of human beings? Walk up to any bird, any reptile, even a wild elephant or lion. How close did you get? Most likely the creature turned tail and fled for the horizon as fast as its legs or wings could carry it, If not, then it was probably cornered, and so, charged. We humans are nearly universally feared by the creatures of the natural world.

It's not fair, you say? You yourself would never harm an animal. I thought so too, until my own secret memory of the kivvers was jarred loose from its moorings that afternoon at Tatsunoshima Bay. And even if there is no blemish upon your own personal relations with the animals, how can you seriously expect a grasshopper to tell the difference between Dick and Jane and DDT Billy? How many years did it take Jane Goodall to thoroughly convince the wild chimpanzees that she was not another poacher with some ingenious long-term plan for their exploitation? Yet how much easier for a *chimpanzee* to identify an individual human than, say, an elk, a garter snake, a bobwhite. We humans tend to say "you beast" as an insult. Undoubtedly, the sea otter, the whooping crane, and hundreds of other endangered species have every right to shout "you human" to describe this "beastliness." And if they cannot shout it, still, they certainly do demonstrate it.

But somehow, and for some inexplicable reason, dolphins are different. Only sometimes do they run away. And even that may be construed as a need to continue on their set course, rather than as a fear mechanism. Play music in the vicinity of a dolphin and there is a strong possibility that the animal will go so far as to alter its set course to venture a bit closer, just to explore the source of the sounds. The wild spinner dolphins off Hawaii went so far as to jump higher and more often when there were humans on the shore applauding their efforts. Yet it nearly strains the limits of common sense to comprehend how such unabashed camaraderie could be the credential of a thinking animal. There have been too many Iki Islands, too many ongoing exercises in human brutality.

Maybe not. Maybe dolphins can discern; and maybe they can see that for every dolphin killer, there are ten or twenty dolphin lovers attempting to make friendly contact. Wade Doak, a dolphin researcher from New Zealand, believes that each and every dolphin may possess an open telepathic connection with all other dolphins around the planet. In a way, it is similar to my childhood idea of a deer's antlers as a kind of TV antenna. Doak has initiated a long-term experiment to test his provocative hypothesis about *collective consciousness.* He has set up an ongoing correspondence with several people around the world who sail the exact type of catamaran as his own. Each boat possesses an identical yin-yang logo on the bow; each participant in the group experiment transmits the same recorded music into the water at predetermined times. The initial results of this study may be read in Doak's book, *Dolphin Dolphin.*

Likewise, author Hank Searles speculates about a worldwide cetacean communication network in his novel *Soundings.* Searles describes the communication network as acoustic in nature. He bases the fact that it is planet-wide in scope on the ability that some species of cetaceans have to transmit sounds for several hundred miles.

The Iki fishermen had their own theory about why the dolphins did not run away from the spears. One of the fishermen commented to me that the dolphin's wisdom focuses upon a loss of fear and a total acceptance of destiny. In fact, the dolphins died just exactly like a Samurai—rushing toward oblivion without the slightest hesitation. The *Hagakure,* a classical Japanese text about Samurai ethics, states it this way:

> The way of the Samurai is found in death. When it comes to either/or there is only the quick choice of death. It is not particularly difficult. Be determined and advance. To say that dying without reaching one's aim is to die a dog's death is the frivolous way of sophisticates. When pressed with the choice of life or death, it is not necessary to gain one's aim.

It was for this reason alone, and not for any of the "Western" reasons like big brains, playfulness, language, or telepathy, that the Iki fishermen came to agree with us that the dolphins were, indeed, special.

The day I saw myself mirrored in the glazed look of a fisherman, I also witnessed a dolphin display this uncanny ability to die gracefully. It happened when the fisherman ripped the net off the dolphin's dorsal fin. Just prior to that event, the same dolphin had been thrashing about in the water, entangling itself more tightly with every movement. But as soon as the animal was dragged bodily up into the boat, it immediately stopped thrashing. Instead, it began to shiver, almost as if it was trying to temper its frantic movements to the new environment of a twenty-foot skiff with four humans aboard. My colleague, Russell Frehling, the acoustic engineer for our project, stepped over the disheveled net and right up to the animal. He began to caress the animal's silky skin, in a compassionate attempt to calm an animal in utter distress. He stroked its chin, talked softly to the animal. The dolphin watched the movement of Russell's hand with its eye. Then the dolphin opened up its mouth. Russell stuck his whole hand into the dolphin's mouth and caressed the gums with his fingers. The rougher fisherman of the two mumbled that Russell ought to be more careful. The animal could easily give a very nasty bite with so many long and sharp teeth.

The strangest thing occurred. The dolphin opened its mouth wider still, and stopped shivering. It stared Russell right in the eye, and seemed to move its head back and forth as if reciprocating the caress. That look, and that action, seemed so totally playful and aware that it gave pause to all of us, both Japanese and Americans. Was this the behavior of a distressed animal? Then the dolphin began to whistle—once, and then a second time. It was a short burst of high-pitched sound, not that different in texture from typical dolphin whistles. It was the context that made it seem so poised. And it cut right through us. Then the dolphin relaxed in

Russell's arms. One of the fishermen responded to this gesture by clapping his hands together, pleased that Russell had calmed the animal enough to allow for an easier removal of the net. The gruffer of the two fishermen, however, was not so amused at this display of compassion and trust. He interpreted the dolphin's gesture as a sign that the animal would no longer thrash about, and so, immediately took action to extricate the net. He grabbed the net firmly in both hands, knocking Russell down in the process. Then he yanked with all his might, pulling the net free, but nearly severing the dolphin's dorsal fin in the process. He was right. The animal did not thrash again, despite the fact that there was blood spurting all over the boat. The animal slipped back into the water without a splash, and sluggishly swam right back into the pocket of the net. It never came up for air again.

Later, Russell and I attempted to analyze this event. At first he expressed a scientific wariness at reading too much anthropomorphic significance into the motivations of the dolphin. However, the two of us agreed that the dolphin had exhibited an almost supernatural steadiness in light of the circumstances. Of course, the dolphin may have been in shock. But then, the animal had communicated a genuine liking and even an *appreciation* for Russell's caresses. The dolphin trusted us completely. We reached the inescapable conclusion that this battered but quite aware animal trusted us humans more than we humans trusted each other. If the Iki fishermen came to compare the dolphins to Samurai, then Russell and I now believed that they were consummate Christians as well.

But if this dolphin did, indeed, trust us completely, the large herds of dolphins out on the banks were also exhibiting first signs of fleeing from the killer boats. Yet only five years earlier, the dolphins displayed only docility at the first sign of a roundup. Six hundred fishing boats encircled the herd. Then a crew member from each boat would lower a long metal pipe into the water. That part of the pipe above water was struck repeatedly with a hammer. It reverberated the loud and sharp white noise through the water. Whether this cumulative din actually hurt the dolphins' sensitive ears, or whether it frightened them, or possibly even disoriented their echolocation ability, has never been determined. Whatever the reason, for five years the act of rounding up dolphins was like corralling cattle, like leading a lamb to the slaughter. Then one day nearly half of a herd of bottlenose dolphins beat the roundup by doing the obvious trick—they swam under the boats and to freedom. That it took even a few of the dolphins so long to figure out this escape has made some Iki observers hypothesize that, of course, the dolphins must have *allowed* themselves to be captured. It was the argument of Samurai courage wed to Christian martyrdom; the dolphins, somehow, must have known that the human media was reporting this ecological atrocity to the world. Maybe their own deaths would save a wolf in Alaska, an elephant in Kenya, etc., and so forth. Those environmentalists who promoted this seemingly farfetched idea believed that this explanation was a lot more reasonable than that other so-called inescapable conclusion that the dolphins didn't know how to escape.

Whatever the reason for the change in dolphin tactics, finally the animals were learning to flee at the first sign of a human being. Perhaps it took the dolphins five years before they could accept the fact that humans were as brutal as they seemed to be acting. Until the Iki incident and possibly a few other widely scattered incidents—such as the U.S. Navy randomly machine-gunning thousands of orcas back in the 1940s, or the establishment of the Turkish dolphin food industry, death by human hands may have seemed somewhat of a novelty. After all, the Greeks and the Romans, probably the Egyptians and the Sumerians as well, treated dolphins as gods. And so, despite the San Diego tuna fishery, the Iki Islanders came to believe that the majority of Americans also viewed the dolphin as a kind of new-age god. There was simply no other explanation for the fact that we made such a big deal over the killing of a few thousand dolphins. And whether or not it was a god in the grand cosmic sense of the word became no more than a question of semantics. If not a god, then a cult, a belief, an object of reverence. Yet whatever any of us may personally believe about calling this quick, cute, smart marine mammal a god, how much better an image than the current Japanese image of raw material for pig feed and fertilizer. And what of the dolphin's image of us? In Japanese waters they are learning what every land animal has already known for ten thousand years—beware the human being. Perhaps there is still some time left to shift that equation. Time for us, as a species, to change our image.

Here we have one of the last animal species on earth to trust human beings, perhaps the animal with the most evolved brain, one of the most dynamic, and yes, one of the animals capable of inspiring us to heights of spiritual discovery; and the Iki Islanders one day decided to publicly slaughter them as a reaction to their declining economy. This, more than anything else, explains the tremendous media interest, the overwhelming guilt that focused on Iki Island. Add to this the fact that a statistical analysis of the fishing business made it only too plain that the dolphins were *scapegoats* for what was essentially a human overfishing problem, and you have created an environmental incident of major international proportions.

The media did its job well to portray the Iki dolphins as our last great animal innocent. The Japanese became guilty of exterminating a kind of cetacean Messiah—possibly one of our betters. Even the Japanese whaling industry came out in favor of the dolphins. There was simply too much cetacean publicity, they said, too much of a runoff onto their maligned business. It is a noteworthy commentary about our animal priorities that the slaughter of hundreds of thousands of starlings in Kentucky by spraying them with a detergent that soaks out their body oils, leaving them unprotected from a freeze, never elicited the rage precipitated by the Iki incident. One wonders, if the buffalo slaughters were now at their height, would there be such an outcry? On the other hand, the slaughter of baby harp seals does hit home with an Iki-style ringing immediacy. Like dolphins, and unlike starlings or even buffalo, baby harp seals are "innocent" and devastatingly cute. Another reason, one that hints at the bigotry, of which the Japanese media reported, is the fact that it is far easier to blame Canadians or the Japanese, than it is to blame our fellow Americans.

Yet there is less violent crime in Tokyo, a metropolitan area of 16 million people, in an entire year, than in any number of U.S. cities during any given week. As a foreigner in such an urban environment, it meant that I always felt safe and secure. Once, I got lost in a strange part of Tokyo at two o'clock in the morning. The man whom I stopped for directions proceeded to flag down a cab, and then directed the cabby to my hotel while paying the fare himself.

What is their secret? How do these people keep their cities peaceful? The answer, if there is a simple one, is actually quite disturbing: Everyone is Japanese. And they are uncommonly socially responsible. They possess a good-natured attitude toward other people, and toward that elusive term known as the human condition. They have had centuries to develop a finely tuned urban protocol, whereas most Americans are still trying to come to terms with the generalities. If the spread-out population of Los Angeles, or even New York for that matter, were somehow condensed to the density of Tokyo, there would conceivably ensue a mass hysteria that would destroy the city within a week's time. In Tokyo, it is business as usual. People smile, chuckle, bow in mutual respect, serve tea, and keep on making babies.

I was sent to Japan by a half-dozen environmental organizations, on a mission to stop the killing of dolphins at Iki Island, and non-incidentally, to promote a more benign attitude toward cetaceans. It was so much more complex a situation than I had bargained for. The very subject of *environment* had developed a negative connotation right across the board of monolithic Japanese society. In fact, "environmentalism," a term that joins natural preservation with politics, was treated as a dirty word. Environmentalists *must* be against the Japanese people because they always wanted to save the lives of those gangster dolphins. Dolphins are the enemy of the fishermen. Thus anyone who wants to save dolphins must also be the enemy of the fishermen. After all, the fishermen do not kill dolphins for sport. They do not even kill dolphins for food. It is a question of outright survival. Dolphins are eating all the Japanese fish. People must eat. Anyone who wants to save dolphins stands in the way of survival for the fishermen who feed the hungry Japanese masses. A dolphin lover is, thus, anti-Japanese.

The Japanese feel that they are unfairly singled out for their rude treatment of old Mother Earth, and many of them resent it. They too have their bird-watching clubs, their environmental impact studies, and their pristine national parks. But environmentalism, as a catchall term for politically expeditious ecology activism, does not meld comfortably with the Japanese manner of effecting change. They simply do not possess the land or the resources to share it all with too many wild animals using too much wilderness. Similarly, they are too dependent upon the sea to share great portions of their precious fish resource with the hundreds of thousands of dolphins who cohabit their coast. How many times did an Iki fisherman say that he wished I could take ten or twenty thousand dolphins back to

California with me? Yet if it were, indeed, possible, how long would that situation endure before the California salmon fishermen began to institute a roundup of their own? Thus the Japanese have tended to view the entire Western concept of environmentalism as a kind of hypocritical idealism. Western environmentalism has also, by now, become irrevocably intertwined with the issue of the killing of whales and dolphins. During the late 1970s, the killings were every bit as much a part of Japan's foreign image as their automobiles and tape recorders.

The 1970s was a time when the Japanese boats used to catch the Japanese fish were built larger and more efficient than ever before. But the depletion of fossil fuels, and especially the Arab oil embargo, made this new fisheries technology economically unfeasible before the boats were even paid for. So, not only did the Japanese begin to catch more fish than ever before, but also, the price per fish spiraled higher and higher. Many previously rich fishing grounds were depleted in just a few years' time. Likewise, more and more fishermen found themselves motoring farther from their home port in search of the large amounts of fish necessary to break even. Many fishermen, usually the ones with the smaller boats and the old technology, went broke. Other fishermen became rich, but only if they continued to expand. So began a quiet panic—polite in the spirit of Japanese manners, but a panic nonetheless.

Consequently, more fishermen congregated at fewer fishing grounds. So it was at the Shichiriga Sone, a still rich fishing bank just ten miles off Iki Island. During the 1970s the number of boats moored at the largest fishing village on Iki rose by 98 percent. The average boat doubled in size from 2.6 tons to 5.8 tons. By 1979, the six square miles of the Shichiriga Sone banks might contain as many as a thousand private fishing boats on any given day.

There were two other problems as well. First, the small bays and estuaries where the food fish bred were showing aggravated signs of deterioration from chemical pollution. The number of food fish—that is, those small fish eaten by the commercially caught fish—began a significant decline. Second, a warm ocean current suddenly began to flow through the area during the winter of 1970–71. Expectedly, the catch of yellowtail, one of the prime commercial species, fell dramatically from 1,228 tons to 452 tons. All these statistics come from reports given to me by the Iki fishing union.

Enter the dolphins. Every winter for possibly many thousands of years, a goodly number of dolphins migrated through the Shichiriga Sone. There were three species. The pseudorca, or false killer whale, is a very large dolphin (up to six meters long) with sharp teeth, a reputation for feeding on other dolphin species, and a history for mass strandings. In October 1946, 835 pseudorcas were stranded on a sandy beach near Buenos Aires, Argentina.

The grampus, or Risso's dolphin, is a pale gray, snub-nosed animal whose diet consists primarily of squid. Pelôrus Jack, the celebrated dolphin who accompanied steamships between the North and South Islands of New Zealand for twenty-four

years between 1888 and 1912, was a grampus. When, in 1904, a man was reported to have taken pot shots at Pelorus Jack, the New Zealand Parliament immediately passed an act granting full protection under the law to any grampus seen accompanying boats through the Cook Straits. Richard Ellis has commented:

> This is generally thought to be the first time that a law was passed to protect an individual wild animal, as opposed to a species, or a particular population.

The third species is the tursiops, the bottlenose dolphin, easily the predominant dolphin species inhabiting the Sea of Japan. They have been seen congregating at the Shichiriga Sone in herds of up to two thousand animals. Ripper was a bottlenose dolphin. The dolphins used in the communication experiments in San Francisco under John Lilly, and in Honolulu under Lou Herman, are all tursiops.

All three species of dolphin consume great quantities of fish and squid. And since so many other fishing grounds around Japan had recently become depleted, it made sense that more and more dolphins would begin to show up at the relatively fecund Shichiriga Sone. So it was. By 1977, the year of the first major dolphin slaughter, fishermen complained that there were "so many dolphins that you could walk to the next island on their backs." That is, of course, if you could see the dolphins for the fishing boats. I find it significant that in my twelve trips out to the Shichiriga Sone, specifically searching for dolphins, I never saw even one.

The fishermen blamed the dolphins for what was a clear-cut case of overfishing. The fishermen also blamed the dolphins despite the fact that the change in water temperature had been locally publicized. The panic had taken hold, and it soon turned ugly. Instead of instituting a management program, the fishermen instituted a dolphin extermination program. It became a case of who would catch *the last* of the fish. And most naively, when the fishermen hit on the solution of killing dolphins, they decided first to alert the media to the event. This single point tells us more about the shortsightedness and lack of worldly sophistication of the Iki Islanders than any other act. The fishermen believed that the media would publicize their great dolphin problem. And that would bring them worldwide sympathy as well as government subsidies. It is true, they did get some subsidies. The government granted them a bounty of $80 for every dolphin corralled, netted, speared, towed, and then ground up in their $100,000 shredder machine. Unfortunately, this gave each fisherman about $28 for three to six long, hardworking days, not to mention costs for fuel. It also kept them from fishing, which was good for the fishing banks, but bad for the economy. But worst of all, the media event netted them international condemnation, embarrassment for their country, and an entire hive full of environmentalists.

Enter the environmentalists. Like the dolphins, there were three species. First were the self-proclaimed dolphin lovers. They arrived on Iki, praising the virtues of dolphins, concocting mini-media events of their own, and offering no practical solutions to the problem. Very few Japanese media people came to these events, no doubt harking back to the national suspicion toward anything smacking of environmentalism. However, the foreign media did come, or at least telephoned from Tokyo. What resulted, after four years of this, was a world situation in which there were more schoolchildren in Kansas and Cologne and Melbourne who loved and revered the trammeled dolphins of Iki Island, than in all of Japan. They do not kill dolphins in Kansas. Subsequently, by 1978, schoolchildren all over the world had learned to consider the Japanese fishermen as savages. I remember Mr. Obata pulling out an entire deskful of children's letters one afternoon. "Why," he asked, "don't they concern themselves with the killing of kangaroos in Melbourne, or the killing of coyotes in Kansas?" I had no answer. Meanwhile, the fishermen themselves bowed at the waist to the dolphin lovers, apologized profusely for continuing to kill dolphins, and then went about their business of organizing ever more efficient roundups that put them deeper in debt. One fisherman was heard to joke, in bad taste, that the influx of environmentalists had spawned a lucrative tourist industry.

Second were the confronters. They condemned the Iki Islanders as "ecovillains," sometimes insulted the union officials to their faces, and once or twice got themselves arrested for physically impeding the roundup process by untying nets and freeing dolphins. Since the fishermen viewed the dolphins the same as they viewed their fish and squid—that is, as property—those who freed dolphins were arrested as thieves, an offense punishable by a jail term. The confronters became heroes in the West, and "gangsters" to the Japanese. Most importantly, the confronters produced very large media events of their own. The issue was finally aired in all its ethical, economic, and ecological glory all over Japan. It made people mad. But it also made them think about what was really going on down there in the rural south of Japan.

Third were the acousticians. They arrived at Iki bearing underwater sound gear with hopes of leading the dolphins, Pied Piper style, away from the banks, or of repelling the dolphins away from the fishing boats. The union officials, no doubt impressed by the science and the technology involved, respected and aided the acousticians up to a point. But when a test failed, such as transmitting the sounds of the orca, an animal who has been known to eat bottlenose dolphins, the negative results were vigorously held up before the noses of the Japanese media as a rationale that continued killings were the only method that worked. Worse still, if the tests worked, such as the initial successes of transmitting loud synthesized ratchet sounds out on the fishing banks, then union sponsorship was rescinded. And without union help, there was simply no other way to get access to the boats, and consequently, no more tests. But why would the Iki union try to impede any method that showed promise at manipulating the movements of dolphins? The reason was twofold. First, many of the fishermen had come to believe that the government would soon step forward with a generous subsidy, permitting the development of a profitable dolphin *industry.* Why else would the government give them a $100,000 shredder? After all, the fish really were disappearing, and a man had to begin to plan for the future. Although most of the Japanese people did not eat dolphin meat, they might learn to like it with the proper promotion. After all, the people of Australia were beginning to develop a taste for kangaroo meat under very similar circumstances.

The second reason involved national pride. Neither the union officials, nor the fishing bureaucrats in Tokyo, could abide the possibility that a foreign environmentalist might actually begin to solve the problem amid what would probably amount to much international fanfare.

During my own two-year Iki career, I either observed or collaborated on or directed all three kinds of environmental activity on the island. I believed that the situation was far beyond any solution that would appease all the parties involved. And each environmental activity, no matter how ludicrous, no matter how devious, offered some valuable lessons on how to go about the business of dealing with an intolerable human/animal territorial conflict. As each event, each confrontation, each experiment, unfolded before my eyes, I became convinced that the continuing media attention was cumulatively positive in favor of the dolphins no matter how it seemed at the moment. And in time, I developed a sincere sympathy for the Iki Islanders stuck smack dab in the middle of an international and environmental conflict that they could never totally comprehend. I knew why they stuck spears into the throats of dolphins. The reason was livelihood. No, not money, not exactly money. Livelihood is money with good, strong, family connections. Entire Iki families were trying to fight their way out of a corner. In their way stood industrialized fishing businessmen, chemical industry businessmen, Tokyo bureaucrats, media people looking for a story, foreign environmentalists telling them how to conduct their fishing affairs, and even a capricious warm ocean current. Not to mention the invading dolphins.

The bottom line for me, as with all the foreigners involved at Iki Island, was the total cessation of the dolphin killings for now and forever. I love dolphins, but I know that I would have felt the same if these had been elephants, or coyotes, or kangaroos. But then, how did I answer Mr. Obata's riddle? I told him that I came to Iki Island because I possessed some acoustic hardware that was relevant only to the specific situation at Iki. It was true. Those had been my ratchet sounds that succeeded in scaring the dolphins. And that day when I watched the fishermen drowning the dolphins, I was also recording their distress signals for later transmission out on the Shichiriga Sone. Evidently this explanation satisfied Mr. Obata. He allowed me to stay in the fishing village while I talked to the Islanders about their problems in a manner similar to a Peace Corps volunteer. Finally, after a confronter, Dexter Cate, untied the fishing nets in March 1980, freeing a hundred dolphins, all of us foreigners were literally thrown off the island. But by then I had accrued enough credibility to compose the first overall analysis of the situation. The paper, which statistically depicted a fishing culture on the brink of economic disaster, with dolphins as scapegoats, was circulated at the various Tokyo ministries involved in the issue. I was accorded press conferences, and attended meetings with the high-level ministers. But of course, the main reason I was given this opportunity to express myself was because Dexter Cate languished in a Japanese jail after perpetrating the largest media event in the history of the issue. And also, because a third environmentalist, Hardy Jones, was circulating a film all over the world that depicted not only the blood of Tatsunoshima Bay, but also the frustration of the panicked fishermen.

This last unified thrust brought the truth of the issue into every home in Japan. This included facts such as:

IKI FISHERMEN LOSE MONEY EVERY TIME THEY KILL DOLPHINS

INTERNATIONAL CRITICISM AGAINST THE KILLING HURTS JAPAN MORE THAN DOLPHINS

ENVIRONMENTALISTS PLAN TO STEP UP PROTEST INCLUDING BOYCOTT OF SONY, JAL, ETC.

DOLPHINS ARE SCAPEGOATS FOR OVERFISHING, POLLUTION, MISPLACED VALUES

Something happened. First the prime minister issued a public statement: "We are asking them [Iki fishermen] to refrain from killing the dolphins for the time being due to the strong repercussions overseas. . . ." Although this statement was made in late March, after the dolphin migrations were already over, its effect remained strong into the next year. Next winter, the killing of dolphins had declined at Iki Island by a very substantial 95 percent.

It is important to add here that there was never any real way to monitor the truth or falsity of that statistic. Another statistic blared out that the fishermen might have already killed one third of the dolphins anyway. Then, later the next winter, I learned that a dolphin kill had been successfully mounted in the Goto Islands, just a hundred miles south of Iki. I had no doubt whatsoever that these were the same herds.

But despite all the twitterings of present and future disaster, all of us "dolphin lovers" had helped the dolphins. We had initiated the slow process of changing the way Japan viewed its dolphins, its fish, and the ocean.

Perhaps by now I have finally counteracted the terrible Karma accrued by my eleven-year-old fisherman self. May I never see another glazed eye as long as I live. May the testimony of this chapter stand as my memorial to the dead kivvers. Like the fishermen of Iki Island, I hang my head and whisper: "We beg the departing spirits to forgive us this terrible deed."

# 4

# Interspecies Protocol

## I. Lions and Bushmen

THE SEARCH FOR NEW ecological metaphors lies at the heart of most nature writing. The case might even be made that unborn generations depend on how willing we are today to explore new modes of perceiving nature. In that spirit, regard the metaphor of interspecies protocol, a concept made of equal parts mythos, Gaia, democracy, animal rights, surrealism, politics, and a sense of place. The concept asks us to treat the animals as peers, neighbors, mentors. There are many precedents from around the world. Consider this tragic account of the death of Bushmen people who lived in the Kalahari Desert.

The waterholes of the Kalahari always served as a magnet for the wild animals that inhabit that part of Africa, including lions, hyenas, snakes, leopards, elephants, and buffalo. Likewise, the lifestyle of the Kalahari Bushmen once revolved around retaining ready access to the few potable waterholes that lay in the region.

The Bushman's oral history focused around the image of human life as it existed around these desert oases for thousands of years. Intriguingly, though the Bushman's stories reveal ongoing instances of tribespeople being mauled, trampled, or impaled by just about every species living there, nowhere in their history is there a single account of a lion killing a Bushman or a Bushman killing a lion.

Western anthropologists visiting Bushman camps during the 1950s often made note of the glowing eyes of the lions, clearly discerned just beyond the reach of the cooking fire. One white observer commented that on those nights when, for whatever reason, the lions started roaring without letup, a Bushman hunter would simply saunter off to the edge of the camp to solicit the lions to please keep the noise down because "there are children trying to sleep." The lions seemed to heed the request. Somehow, both species had long ago developed a protocol for living at peace with one another.

The traditional Bushman lifestyle ended forever with the introduction of ranching into the Kalahari during the 1950s. The waterhole culture deteriorated to make way for Westernization, including such detritus of progress as the introduction of rifles, four-wheel-drive vehicles, and an externalized economic system based upon raising cattle for money. Significantly, the Bushman/lion protocol was soon replaced by a mutual attitude of disrespect and fear.

Before the introduction of ranching, Bushmen and lions kept very strict schedules about when and when not to visit the waterhole to drink. Lions drank late at night; Bushmen filled their gourds during the heat of the day. Surprise encounters were kept to a minimum by a strict adherence to what was a de facto scheduling protocol. Ranchers were, of course, oblivious to the protocol, and so cattle started appearing at the waterholes at all hours of the day and, especially, at night. At first the lions actually seemed to keep their distance from the cattle, as if this new species was to be respected as a living "extension" of the human community. But as traditional schedules became ever more disturbed, lions finally started attacking the easy prey. The ranchers, with help from their Bushmen hired hands, soon reciprocated by shooting lions. Ironically, within a few years' time, several Bushmen had been killed by lions. Now, just twenty years later, there are no Bushmen living the traditional lifestyle. And no interspecies protocol.

## II. Clownfish and Anemone

Wander out into a cow pasture on a hot summer day to observe cattle grazing in the midst of egrets or starlings. How close did you get before the first animal changed to a defensive posture? How close did you get before the last bird flew away? If interspecies protocol were ever to become a science, then these two distances might provide a potential statistical baseline of data. Simply put, the numbers reveal the distance at which trust breaks down between your human self and your wild neighbors.

Webster defines protocol as the ceremonial forms and courtesies that are established as proper and correct in official relations between parties. Interspecies protocol may thus be understood to mean the forms and manners (and defense postures) that any species conforms to when relating to another species. Such protocol might be based upon a species-specific instinctual behavior, or in certain other instances upon learned behavior agreed upon by individual animals from two or more different species. The classic symbiotic relationship between clownfish and anemone offers an instance of the former. Sea anemones offer a significant protection for clownfish within their stinging tentacles, but do not sting them. Clownfish are poor swimmers, vulnerable to predators when not in the protection of their host anemone. Clownfish also lay their eggs at the base of the anemone. In return, clownfish usually consume their prey within the anemone's tentacles, in which case the anemone consumes the leftovers.

The historical relationship between lion and Bushman is an example of the latter.

Protocol is different from symbiosis. Symbiosis implies a physical co-dependency advantageous to two dissimilar species. Protocol is, instead, a social behavior established between individuals. Regard it as the etiquette of an ecosystem. It brings to bear such concepts as demeanor, posture, and even ritualized interaction as a means of communicating and building trust between species. Because protocol inevitably implies varying degrees of conscious behavior communicated between species, it may be unfairly construed to be anthropomorphic, which is the precise reason we never encounter the concept within the standard lexicon of field biology. But who can honestly say where instinct ends, and a mutual conscious regard between individuals begins? Is the relationship between clownfish and anemone entirely instinctual, and therefore generic? Or do individual clownfish and individual anemone also need to fine-tune their obviously symbiotic relationship to best fit one another's precise size, habitat, level of trust, and even personality? If this occurs (and why not?), then protocol is the term for it. Symbiosis thus refers primarily to relations between species. Protocol refers to relations between individuals.

How is an understanding of such protocol in nature important to us? In fact, when we forget that we are interdependent with any (or all) other species, those other species suffer. We suffer as well. Look around. Who can deny that nature is everywhere in retreat? This cessation of protocol may eventually spell the death of humanity. Significantly, if a modern human being is to accept the concept of protocol, he or she must also accept the idea that animals are possessed of individuality and distinct personalities.

Permit another example. Observers of the social behavior of predator and prey on Africa's Serengeti Plain comment that wildebeest often display an astonishing lack of fear of lions not in hunting mode. Although the deportment of lions at rest is generally recognized to be a subject transmitted by adult wildebeests to their young, turn the idea upside down and you may realize that the lions themselves probably play some significant part in the transmitted lesson. If so, then it implies a shared sense of community between lions and wildebeest. A protocol.

## III. The Most Fearful Survive

The word *protocol* originates from two Greek words: *protos* (first or primary) and *kolla* (glue). In medieval literature a protocol was the term used for a table of contents—the first page glued into a manuscript cover. In that sense, regard interspecies protocol to be the consciousness, or glue, attaching an individual to its social environment. A strong protocol implies well-developed bonds of communication between species, while a weak protocol is essentially distrustful, or perhaps no relationship at all.

Protocol may develop into such a formalized behavior that it appears instinctual (symbiotic), although when it is seen through fresh eyes we may just as easily perceive it as a set of rules passed down through generations. Grizzly bears in northern Canada have always been routinely shot on sight. Over hundreds of years, the bears who ventured closest to human beings, for whatever reason, were the ones who were shot first. Consequently, those individual bears exhibiting any measure of outgoing curiosity that might have conceivably developed into a strong protocol were the first to be done in. Only the stealthiest and most fearful of the species survived. Over time the bears became what we made of them. They co-evolved into creatures instinctively fearful of, or angry with, human beings. We live with that legacy of weak protocol today.

## IV. Language Reflects Perception

We modern human beings cannot properly comprehend the ramifications of interspecies protocol until we first learn to perceive the other animals in an entirely new way: as individuals, and possibly, as peers. We need to know the lions as the Bushman knew them. Interspecies protocol helps us examine our inability to relate to animals as a kind of bad politics. Social relations void of compassion, without mutual trust and mutual respect.

To understand protocol we must learn to speak about nature in a way beyond contemporary language. Psychologists tell us that language reflects perception, leading to a certain worldview. For example, the prevailing educational system teaches our children to preclude words of feeling and equality—words like intuition, love, participation, magic, and communion—whenever they draw close to observe wild animals. Or another example: Understanding the marriage of perception and language sheds new light on governmental wilderness policies molded by urbane and socially polished men of power debating legislation in a linear fashion inside artificially lit rooms. These are mostly well-educated people who, like the rest of us, believe that "natural resources" is a valid synonym for wildness. Yet whereas wildness was once central to any human being's daily perception of the world, now it is reduced to the margins of our lives. Our language and our policies mirror that distancing. We get the policy we perceive.

By contrast, that most intuitive and non-linear of books, the *Tao Te Ching*, offers one of the oldest as well as one of the most succinct treatises about a developed protocol between human beings and nature. As the *Tao* says: the relationship with nature that can be defined is never the real relationship with nature.

The Colville Indians, who once inhabited the shoreline of the Columbia River, tell us the story of the river monster, Nashlah, who has been eating all the people (meaning both human and animal) who travel up and down the river in their canoes. The trickster, Coyote, comes to the rescue by allowing himself to be swallowed into the belly of the monster. He lights a bonfire out of the jetsam left by former victims, while proceeding to cut up the monster's heart to both warm and feed all the half-dead animals inside. As the monster grows increasingly weaker from his wounds, he starts coughing up all the animals, who are thus saved.

Then a plot twist. Unlike most archetypes of resurrection, for instance the story of Pinocchio in which the evil whale is slain, the Colville assert that this is just the start, and not the conclusion, of a long-term relationship between a sea monster and Coyote. These two are after all not enemies, but neighbors bound by a common search for a unifying protocol. The second half of the Colville myth recounts the resurrection of this monster, Nashlah, who is now given a strict admonition against killing every person who travels the river. Even the most dangerous predator is accorded status within the status quo, and so deserves certain rights to live and enjoy good health. In this instance, a protocol is negotiated between Coyote and Nashlah. Our monster neighbor is permitted to shake and overturn only those canoes that pass directly overhead. With this last image lodged firmly in our minds, we finally understand the beast, Nashlah, to be the mythological keeper of an actual rapids that flows along the Columbia River.

But what Coyote would never do, the Army Corps of Engineers has accomplished. A recent dam built on the Columbia River destroyed the Nashlah rapids. Today, all that is left of the monster is this myth. Had the engineers taken the myth more seriously in their rush to better control the river, we might have been led to a protocol about how to live with this river. In this case, it would have saved the salmon that are now endangered all over the Pacific Northwest because of too much irresponsible dam building.

## V. Who Are the People?

Once again, the use of words like *people* and *neighborhood* to describe ecosystems underscores the fact that language mirrors worldview. And whereas we Westerners learn most of our right relations between people from the texts of our Judaeo-Christian heritage, in reality, the exclusivity of this heritage also articulates the foundation of the current environmental crisis. Jesus, for example, asked us to love our neighbors (other members of the human species) as ourselves. By contrast, the Colville myth strips Jesus' aphorism of all its specieist inference, now including the ecosystem as the very foundation of neighborhood. Or attach a Colville spin to Abraham Lincoln's democratic construct about a government of the people, for the people, and by the people, and you soon find it transformed into a quite disorienting statement binding democracy to ecology and evolution.

So hundreds of traditional myths would just as soon strip Lincoln's aphorism of all its specieist inference, and include all species/people as the foundation of citizenship, with relations between peoples governed by interspecies protocol. In this version of government, it is the ecosystem, the watershed, the bioregion, rather than the nation-state that becomes the wellspring of political consciousness. This expanded metaphor also offers a faraway glimmer of a new kind of democracy. It is a democracy that equates a sense of place with community. It is a democracy based on interspecies protocols that once upon a time defined harmonious relations between species. Gary Snyder puts the idealist tenet this way:

> What we must do is incorporate the other people . . . the creeping people, and the standing people, and the flying people and the swimming people . . . into the councils of government.

Atavistic? Yes, definitely, when used in the context of governmental structures as they exist during these days of accelerating environmental degradation. But given the ominous reality that faces the planet, human beings everywhere must start to acknowledge other ideas capable of being persuaded onto the table of possibilities.

# 5

# *The Rite of Passage*

## I. Adan

Adan was a little boy who lived with his family in Dallas, Texas. He had been born with a severe heart defect, and thus had never been able to synchronize the little boy's enthusiasm with the old man's tired body. However, recent surgery had been deemed a success. For the next full year Adan was happily active like any growing boy. It lasted one full year. Then, one winter afternoon shortly after his sixth birthday, Adan suddenly died as he played about under his family's dining room table. His body was cremated. The ashes were packaged inside a classic-style Greek urn, to be buried or dispersed as the family saw fit.

Adan had loved whales. He had often spoken about sailing out onto the deep blue sea to jump into the water and play with all the whales who dropped by. It was something that he knew he had to do.

His mother, Denise, decided that it would be a meaningful tribute to her little boy's memory to carry Adan's ashes to some ocean, any ocean, and then scatter them about. Maybe whales would appear, who could tell? Denise telephoned her good friend China, who lived in a big house overlooking the Pacific Ocean in northern California. Together, they discussed the death, the cremation, the urn, the fantasies. Denise asked if a boat might be found. The ashes would be cast to the four winds. They would conduct a "rite of passage" for Adan. Somehow, they would invent a service that would assure that Adan's soul would travel securely to heaven. Or to Nirvana, or to reincarnation, or to wherever souls choose to go. Neither woman wanted anything pretentious—in a way, they both sought a "going-away party" with more heart, more soul than a typical funeral.

China agreed to locate the boat. She then mentioned to Denise that there was a man who had lately been receiving a lot of publicity for his ability to attract whales by playing music to them. Denise agreed with China that it would be very worthwhile to find this man. Here was an opportunity to combine her son's funeral service with real live spouting whales. Adan would have been tickled pink. And that was good enough for Denise.

## II. The Whalesinger

It is invigorating floating around in the open North Pacific Ocean. I am thirty yards from the rubber boat, lying flat out on my back; rubbing this carved "whalesinger drum," this interspecies musical instrument, descendant of the Native American tone drum, the tulke of the Maya, the tepanatzli of the Aztec. The sound is generated by rubbing the various wooden surfaces of the box-shaped drum with a mallet tipped with a dime-store superball: something like scratching chalk against a blackboard, but much more mellifluous. Different-sized superballs emit different pitches, textures, and octaves of sound. The giant red, white, and blue superball moans ever so deeply, and sometimes clicks as well, if it is bounced and rubbed at the same time. The technique for this is quite tricky, and it took several days before I had mastered it. The medium-sized cat's eye superball sounds roughly akin to the human voice. But it is a voice of rubber and is ever so long drawn out, like a whale's. A whale human. The small Day-Glo orange superball screeches like an elephant, but not quite so harshly. All these sounds, taken together, evoke any and all of the whalesongs that you or I have ever heard; either in reality, or in the deepest recesses of our animal dreams.

The whalesinger has been equipped with an outrigger, which keeps the instrument from tipping and drawing water through the tuning slits that have been cut into the top face of the drum. It also gives me a snug little frame to put my body inside of, and thus be totally supported while riding through the swells. This is more than a case of becoming one with the instrument. The swells can get as large as twenty feet. Interestingly enough, it is not the tops of the waves that bother me. Rather, it is the valleys, between the swells, when I cannot tell for the life of me where either the boat or the shore has gone to. I am not out here to test the hydrodynamics of slit drums. Neither am I out here because I enjoy three-hour immersions in 42-degree water. Rather, I am here because I believe that humans can communicate to whales.

At this point in the long-term venture of interspecies communication, it seems essential that I play this music from directly within the whale's watery environment. So here I float, a mile offshore of the rugged Point Reyes Headland, working up a musical sweat, making whale-type sounds with all my body and soul. And the gray whales, these forty-five-foot living express trains, are all about as they continue along their leisurely four-thousand-mile swim from northern Alaska to the warm harbors of Baja California. There they spend a toasty two months: congregating, courting, making love, and making babies. Once the babies are born and taught a bit of practical seaworthiness, the grays turn about and immediately head back up the coast to northern Alaska again.

Scientists say that whales swim sixty to seventy miles a day. The entire journey takes about three months. While they are migrating, whales will surface and breathe two or three times at ten-second intervals. Then they throw their flukes high up into the air and submerge for three to five minutes. The gray whales used to also live along the coast of western Europe. Likewise, there was another gray whale migration along the Pacific coast of Asia. Both populations were exterminated by whalers. In fact, our own West Coast population nearly succumbed to the whalers as well. But laws were passed just in time, and the pods began to make a comeback. Today, all fourteen thousand known gray whales migrate up and down along the western coast of the United States, usually swimming less than twenty miles from shore. Sometimes, they come so close to shore that it makes you want to get out and swim with them.

Watching the spectacle of this migration has developed into quite an environmental event at various headlands along the coast. There are incredible moments when one is able to see a pod of twenty or thirty whales, all less than a half mile offshore, spouting their breath ten feet into the air and occasionally lifting their enormous leviathan bulk completely out of the water.

Former Governor Jerry Brown of California was only too aware that the whales' migration along the coast was a kind of natural wonder on a level with the redwood trees and Yosemite Valley. To publicize both the migration and the plight of the whales worldwide, Governor Brown sponsored "California Celebrates the Whale." First, there was the weekend event in Sacramento, with lots of music, exhibits, and movies. The event gathered most of the professional whale-saving human beings from all over the planet under one roof. In retrospect, I always felt that the celebration was the first governmentally sponsored totem event in white United States history. As the environmental philosopher Ponderosa Pine has said about the event:

> We learn to sing and to dance as a way to provide ourselves with our own medicine, and eventually feel as good as the porpoises do.

The event was such a success on all fronts that the governor decided to continue his active promotion of the whales. The state hired a group of innovative design consultants, collectively known as the Ant Farm, to design some program that would bring the event of the gray whale migration closer to the people. The Ant Farm, best known for such arty-type works as the Cadillac Ranch (a monument to the tailfin) and the House of the Century, which looked vaguely like an electric shaver, went to work. They designed and built an underwater public address system that would make the sounds of the gray whales audible to listeners on shore. The hardware consisted of a buoy containing underwater microphones known as hydrophones, all connected to a radio transmitter. The signal could then be received by any radio with a public service band, and from as far away as two miles.

Unfortunately, the gray whale is known to be one of the less vocal whales; most of their vocalizations occur in the harbors of Baja and in the waters off Alaska—in other words, anytime when they are not migrating. Thomas Poulter, a bioacoustician who had recorded over sixty thousand feet of tape of gray whale sounds, categorized their vocalizations as clicks, rasps, and the bong of a big Chinese gong. If The Ant Farm's project were going to succeed, they would have to devise some technique for coaxing the whales to sing. Since I was enjoying a modicum of success in my playing with dolphins, The Ant Farm hired me as their musical director. I was to wield the baton. My chorale easily outweighed the Mormon Tabernacle Choir. Nine times during the months of December and January in 1976–1977 I slipped into my wetsuit, draped my arms and legs around the whalesinger drum, and slipped into the waters of the Pacific Ocean. The music that I made on the drum resonated out through the bottom of the instrument and thus directly into the water. On several occasions it attracted the attention of passing gray whales. Sometimes they came so close that I felt as if I could reach out and shake their hand. It had something to do with their eyes.

## III. Herbal Poetry

Denise and China moved ahead and rented a sailboat. I was hired as interspecies musician. They both wanted to know if I could really attract a whale to the boat. I promised nothing, adding that it was quite impossible to predict the behavior of these forty-foot sea mammals. "But," I added, "if everyone involved in this event is focused around the ideal of 'contact,' then all of our collective transmitters, emotions, our bodily receivers will be that much more in touch with the known and unknown powers that be." A ceremony of death may very well be the most sharply focused event that we humans can muster. Grief discourages pretension. The rite of passage seemed a potent interspecies experiment in group telepathy. If the whales could indeed feel human vibration as so many seem to believe, then this funeral service would probably provide the optimum occasion for plumbing the depths of that relationship.

There is a species of dolphin, the pseudorca or "false killer whale," that has been documented holding its own funeral service. The event was filmed off the Dry Tortugas, west of Florida. The entire pseudorca family had followed their dying pod leader right up onto a beach. They then proceeded to buoy his pained body with their own, in an attempt to keep his internal organs suspended off the bottom, and to keep his blowhole right side up. There they all remained for three days, taking turns alongside the aged bull, whose condition continued to deteriorate. Finally he died. And it was at that moment that the rest of the family chose to turn about, heading out of the shallows back out into the open ocean.

China wanted to know what our chances were of even spotting a gray whale, let alone communicating to it. The grays had begun to leave their winter calving grounds in the warm Baja lagoons. A few had already been sighted locally, midway in their long migration. However, all of my experience with grays had been accumulated during the southern migration of early winter. During the peak of this run, in late December, three or four hundred whales might pass during a single day. But the northern run is much more spread out over four, five, or six months. It is rare to see more than a hundred pass any point during any twenty-four-hour period. But for our purpose, this northern migration had one major advantage. The whales possess no birthing urgency. There is no longer any rush to get south. Instead, they tend to swim leisurely, pausing to linger wherever something of interest catches their fancy. They have been known to swim around a boat for hours at a time. Gray whales have been documented picking a small skiff up on their backs, carrying it a short distance, and then setting it back into the water. The peak of this leisurely northern migration occurs at the time of the spring equinox. That was only a few weeks away. But should we set sail on the last day of winter, or the first day of spring?

The weather decided for us. Saturday the twentieth of March was gusty. By midmorning it had turned into a small crafts advisory. So on a very early Sunday morning in March, the first day of spring, our group sailed beneath the span of the Golden Gate Bridge and out into the Pacific Ocean. The group consisted of Denise, China with two of her children, boat skipper Tom, filmmaker Will, who had been invited to document the strange event, and myself with whalesinger in tow.

During the long passage out to the whale grounds, Denise sat by herself on the stern, her sad eyes protected behind enormous sunglasses. She clutched the urn to her bosom as if to share her body warmth with Adan for the last time forever.

Just last night, as we sat around waiting to hear the weather report, Denise had expressed a confused, last-minute apprehension over the dignity of this "California style" event. She questioned each of us in turn; trying to get to know us, yet also seeking some hidden answer to the mystery of death that had drawn us all together. There were many tears shared among friends. I finally made it to bed, totally bewildered by this concept we call death. The very idea of "foreverness" yanks us away from our comfortable lifestyles, out from underneath our secure self-image. Instead, it places us smack dab in the middle of the universe. We arrive confused, unable to plumb the depths of this unfamiliar infinity, this eternity. We have no handle, no substance upon which to hold. A confrontation with death becomes a trip into a conceptual isolation tank. And just at that moment when our senses no longer suffice to provide us with firm ground, the door of our mind opens up just a wee bit deeper into its own hidden potential. In a way this metaphysical opening makes us more available to the otherwise hidden mind-to-mind relationship with the whales.

I have chosen to sit by myself in the forecabin of the sailboat. The whalesinger sits disassembled in a heap on the bunk across from me. For the occasion I have brought along a paper bag full of what I choose to call "herbal poetry." Its contents lie on a clean white plate on my lap. First a raw chunk of elephant garlic is diced up and dropped into a glass jar. Garlic obviates the ill effects arising from a sudden change from air to water. Next, a generous slice of ginger root is cut up and stuffed into the jar. Ginger enhances the clarity of the brain. Now a bit of dulse, a purple seaweed with a strong taste of the ocean. Finally, three mushrooms are diced and dropped into the jar—a symbol, a smell, and a taste of the earth. I scoop up some seawater through the porthole and place the mixture next to the space heater. Heat will allow the ingredients in the potion to marry. The resultant brew should be a real kicker, one more way to clear our minds of all the day-to-day emotional garbage. This day is to be special. I have decided to give the occasion my best shot. There are only thirty whales a day swimming across a path that is eighteen miles wide. We are going to need all the help we can get.

## IV. Omatakwiase

During December, out with the gray whales for days at a time, I opened myself up to many strange and varied techniques for successful whale communing. I consciously became a kind of interspecies guinea pig. The word got out very quickly that I was an amateur shaman on the lookout for strong yet practical magic. An influential friend convinced me to investigate the power of mantras. One repeats a special word or seed syllable over and over again until the vibration generated by the word sinks into one's very essence. OM is a well-known mantra. HUM is another. However, I had to admit that all the prescribed Sanskrit mantras only made me feel self-conscious. I always felt like the cowardly lion in *The Wizard of Oz* repeating: "Oh I do, I do, I do believe in spooks." Then another friend took me to an American Indian ceremony where I heard the word *Omatakwiase* repeated over and over again. I loved the way the syllables rolled off my tongue: *Oma tak wee a say.* The word comes from the Dakota language and means: "to all my relations." The phrase is used as an analog to our own "amen." Traditionally, chanters offer prayers to the heavenly grandfather. At the end of each invocation, the speaker chants: *Omatakwiase.* The word is imbued with a power to bring the speaker from the heavens and back down to earth again. The Sioux say that this word is not to be spoken lightly. Intuitively, the sound symbol seems to focus its speaker back into the earth and to all the creatures. It is a phrase of totem.

I started repeating the word, rolling over and under the swells, saying it over and over again. As a practical measure it worked to center my mind in one-pointedness, and away from all the mundane concerns of my chosen work, such as the unpleasant sensation of cold seawater coursing through an ill-fitting wetsuit.

Next, I learned a new way to breathe. I began scooping in the air like any normal sea mammal. I read about the Tibetan monks who sit out on the midwinter lake ice, naked. A practice they call "firebreathing." Breathing fast and furiously, they bring in and expel the air in such a way as to generate enough body heat to melt clear through the ice and so, fall into the freezing waters below. An appropriate punch line to one version of the story goes: "And aaaaaah, what a refreshing dip it is."

But by far the most captivating sensation of this, my period of gray whale apprenticeship, was the weightlessness. I am more than buoyant in this spongy wetsuit with the classic orange life jacket clinging to my shoulders. The balance between gravity and buoyancy is just about equal. This feeling of suspension begins to do absolutely bizarre things to my brain. Sometimes, if I keep my eyes closed long enough, it becomes impossible to perceive if I am all mind or only mind. The first way includes my body in the total sensation. The second evokes an out-of-body experience.

Immersed within this sensation, it becomes very easy to stop rubbing the surface of the drum. This is obviously dangerous. Can you imagine if I hypnotized myself while rolling around in huge Pacific winter swells? So I force myself to rub the drum. The vibrations course clear through my body and so enter the water. The instrument vibrates so intensely that every once in a while it splatters water off its surface with a violence that sends the spray a good six inches into the air.

And once in a while, a gray whale will approach quite closely to check out the source of the sound. There is no warning. Suddenly, from seemingly out of the cosmos, a huge shape rolls across the surface and explodes a fishy fragrance into my face. When this meeting occurs at the crest of a wave, the sensation is similar to encountering a fellow sightseer while inspecting some remarkable vista. We stare intently into one another's eyes. However, when we meet in the canyon between swells, the feeling is more akin to two bodies sharing a watery sleeping bag. I am Moses out in the middle of the parted Red Sea with frothing walls of vertical water on each side. These walls deaden the sound of the wind far better than any boat. The silence screams. The whale's blow resounds. For a brief second the two of us share a small room. This time the eye contact feels so much more intimate. Then the room is turned inside out and the whale is gone. I am thrust back onto the summit, where I catch a glimpse of the departing spout.

Fifteen minutes later, I suddenly wake up to realize that my mind is drifting again. And so the drumming begins in earnest. The Ant Farm people, watching me through binoculars, begin to realize the potential danger in these forays. So the next time out they insist on attaching a hundred yards of line to my foot and to the boat. This umbilical serves to make it that much easier to waft away into trance.

But do not be misled into a presumption that I am simply daydreaming out there. No, the overall sensation is one of extraordinary consciousness. I have never felt quite so awake, quite so immersed in the slow and slower and slowest of motion. Tied to the human race by a flimsy yellow nylon cord, I have never before felt so radiant as during those precious hours spent in the presence of the gray whales.

On the last day of the radio project, I finally achieved what might be called a communication breakthrough. Typically, I was lolling within the swells—fire-breathing, chanting, weightless, whalesinging, warm, and happy. An hour previ-

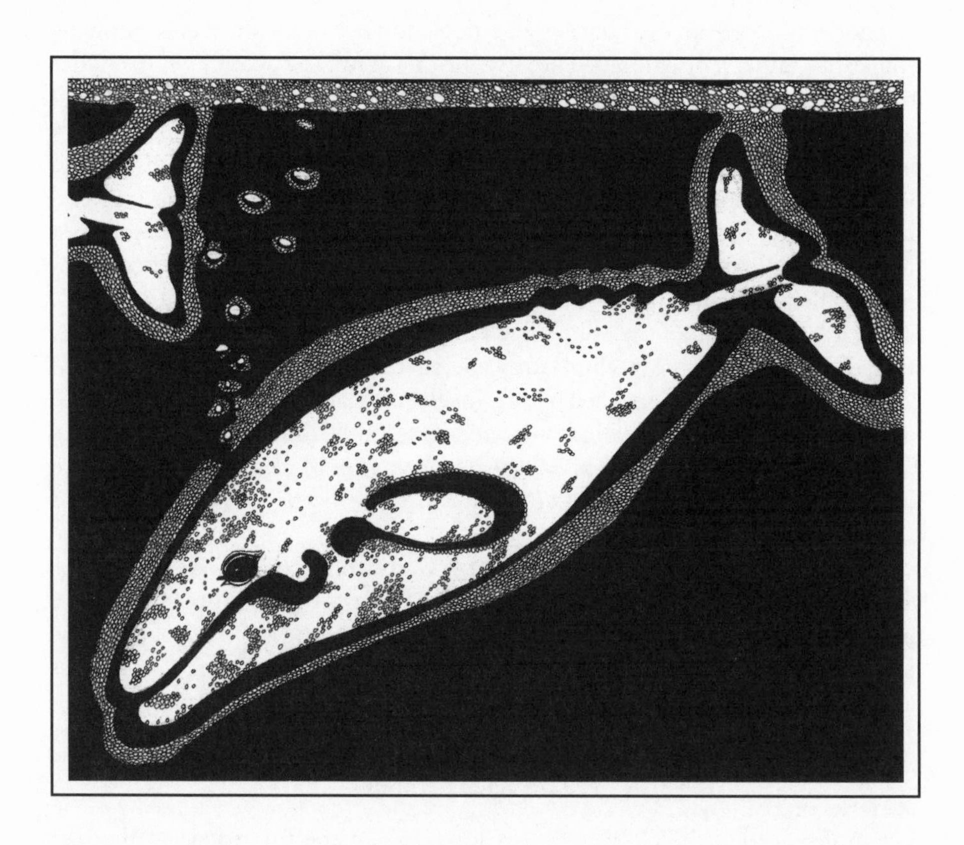

ously a gray whale had swooped in for a closer look. The whale had gaped open its mouth as if in anticipation of a repeat of the Jonah story. My own mouth must have gaped open as I stared at the improbably large mouth filled with baleen. Gray whales do not have teeth, but rather a kind of comb material. They feed a little bit like a backhoe, sieving tons of mud and water while successfully entrapping vast quantities of clams and crabs. Then the whale closes its mouth and slips beneath the surface.

Suddenly I "see" an eye above my head. Not an eye in the head of a surfacing whale, but rather a disembodied eye like the one above the pyramid in the dollar bill. To be precise, the perception was similar to the sensation that forces a person to turn around when someone is staring at him. There was more to it than that because it was not only a sensation but also an eye. It was dark brown, cradled within the folds of a heavy skin. At the edges, where the rest of the face should have been, the eye metamorphosed into the intense blue of the midday sky. At first the eye floated above my head, peering directly down. I finally comprehended that I was not truly "looking" at it, because it moved behind my head. Then, the perception of the eye plunged into my skull. I felt it moving first here, then there, surveying the cerebral landscape, surmising the territory. I felt my own mind in a way never experienced before. Inside my head existed a vast unexplored cavern cross-gridded with an immense network of side tunnels. And all the while, this disconnected yet conscious eye was walking around like a spelunker out to document the handiwork for both of us. I was able to watch both the eye and also what it saw.

Then a whale surfaced no more than five yards from me. We were both in the trough of the wave, like two peas in a pod. Or to be precise, like a beach ball and a pea in a pod. The whale blew with the reverberation that is so distinct to all the large cetaceans, and started its dive. It lifted its flukes high into the air above me as a sign that this time it was leaving the area. Just as suddenly, the eye disappeared.

## V. The Bass Canary

Five miles beyond the Golden Gate, China sighted the first whale of the day. "Hello, hello," we all shouted in unison. But this gray whale was easily a half mile off the bow, plying north in the morning sun. We stared in anticipation. But by the time we had reached the general vicinity, the whale was gone.

Eleven A.M. Fourteen miles off the mountainous California coast. The boat rocked back and forth, going nowhere. The Farallones Islands, home to hundreds of thousands of seabirds, loomed ominously above the fog, six miles to the northeast. Skipper Tom sat at the helm, spinning a sea yarn about the time he encountered forty-foot seas off the south coast of Tasmania. China's son Ben, Adan's former playmate, stared up at Tom in rapt attention. Will, camera in hand, moved about the deck, running off roll after roll of film. China lounged on the foredeck, soaking up the rays of the hot sun. Madeline, China's daughter, sat forlorn in the forecastle, exhibiting signs of impending seasickness. Denise sat alone at the stern. Finally she got up from her nest, and climbed below deck to store the urn. At best, she had half listened to Tom's story. She remained the mother in mourning. One needed merely to look in her direction to replenish one's sense of mission.

I was seasick. I had already vomited four times, feeling drained, out of control, a bit embarrassed. I felt like a shaman turned wimp. Just a few days before the trip, I had discussed my sense of readiness to meet the whales during this rite of passage. A friend recommended, "Hope to hell that you vomit. That's the best cleanser of all." So there I sat in the cabin, ready to be sick once more, watching the cups roll on their hooks. Denise came below. I did my best to appear composed. She looked past me, and beat a hasty path to the toilet. Her face was as green as a seedless grape. Suddenly there was no longer any need for embarrassment.

Another hour passed. I felt much better by now, venturing out into the open air. Up on the bow, Will gestured excitedly off to starboard. Yes, we all saw it—a spout less than a half mile away, seeming to head our way. Soon, it became obvious that this whale would pass just to the west of us. Tom quickly set sail and began to follow in the path of the whale. Slowly I pulled on my wetsuit. Next, I assembled the whalesinger, and hauled it up to the gunwales in anticipation of easing it down to the water. There, on the side of the drum that faced me while I floated, a tiny school photo of Adan stared out with a little boy smile. Yesterday, when I had asked Denise for the photo, she had hesitated, and then questioned me about my intentions.

I patiently explained to her all my various methods for achieving a certain level of one-pointedness: the mantra, the breathing, weightlessness. I told her the story of the eye walking around inside my head. I did not ask her to agree with my perception of what had happened. I asked only that she acknowledge the possibility of mystery. We discussed telepathy. It may best be described as simply a kind of "bioradio." Sometimes, when the radio is on, the reception is strong, we are able to listen to beautiful music. Other times, no matter how hard we try, all we hear is static. Some of us are able to transmit as well as receive. However, unless we are familiar with the code, it all sounds like nothing more than a lot of dots and dashes. There is no earthly reason for us to assume that such abilities begin and end with human beings. In fact, it might be true that a whale, with its large brain, could conceivably possess more wattage than a human's bioradio. If we acknowledge even the possibility of interspecies telepathy, then it made good sense for me to be focusing on Adan's likeness as I played music in the water.

Denise handed me the photo. That night I glued it onto the front face of the drum.

I stood poised on the afterdeck, drum balanced on the gunwale. The entire group looked at my face for direction. I opened my potion of herbs, which lay on a pile of ropes, and begin to spoon it into my mouth. I explained all the ingredients of the brew, and asked that we share a taste as an earth Eucharist. The sentiment of sacrament, albeit improvised, brings disparate emotions together. Tom, Will, and China each took a sip. Denise demurred. She stammered, and then tried to explain that it made no sense to her. No one insisted. I explained that it was a simple attempt on my part to share a bit of herbal poetry. She liked it when it was explained that way. Finally a twinkle crossed her face, and she took a sip.

I quickly descended over the side of the boat. Tom had tied a goodly length of line to my ankle, and it pulled me off balance as I hit the water. After a few seconds of adjustment, I finally managed to wrap myself around the drum's outriggers. Immediately, I started to tap out a clear moaning rhythm. There was nothing else left to do but go to work; pound out the *waaaaaaaaaawooooooooooopity-thud-thud* in a catchy syncopated beat and so get my blood circulating. The water was very cold.

But as always, the doubts slowly began to loom into prominence. What was I really trying to do in this deep water? Play music with a whale? Most people would have a hard time understanding this *waaaaaaaawoooooopity-thud-thud* as music at all. But I like it. It is loud, good, strong drumming. I have been listening to a lot of Bob Marley music lately. That strange reggae offbeat inflection has not been lost on me. Do whales like reggae? Or for that matter, can this particular whale hear it at all? Yes, it is certainly loud enough. A whale possesses an ear like a bloodhound's nose. *Waaaaaaaawoooooopity-thud-thud, waaaaaaaaaawoooooopity-thud-thud.* Do it boy! All at once I begin to warm up. My stomach clicks into its contented gear. I slowly begin to alter the rhythm, spacing the individual riffs at twenty-second intervals. I stare at the photo of Adan, and very quietly chant *Omatakwiase* in a deep voice. Fifteen minutes pass.

Suddenly, some very startling sound issues from right out of the water. At first I cannot quite believe that I have heard it. Is it a bird? No, there are no birds in sight. Is it a plane? I do not see any. It sounds like a bass canary recorded at 78 RPM and then played back at 33⅓. I hadn't made the sound; the drum is simply not that deeply resonant. At the crest of the next wave I notice that the boat is at least two hundred yards away, so it wasn't that either. Then, as if in answer to my questions, the sound erupted again, identical in phrasing to the first sound. It is not a rasp, a click, or even the bong of a Chinese gong. No, this is more like the evocative song of the humpback whale. The major difference is that perhaps this song is terser than a humpback's. I recall a conversation with a tour captain from Scammon's Lagoon, one of the Baja birthing grounds of the gray. This man had heard sounds that reminded him of whale songs, more melodic than anything the scientists gave the grays credit for. Whatever the truth may be, at this moment I am hearing some being singing a song in my vicinity. It deserves a reply. Strangely enough, I begin to howl like a wolf.

I play and chant and howl for about ten more minutes. Ten minutes of lying very still in the cold water. But my concentration has been broken. I have been nonplussed by an unidentifiable bass canary. My body begins to shake and shiver. I lift up an arm to signal the boat. A minute later I am dragged aboard, the drum left dangling over the side of the gunwale.

I beat a hasty retreat to the forecabin and huddle in front of the tiny space heater. I look up; China is standing in the doorway, staring at me with a huge smile on her face. Had she heard the sound? No, only the drum. Then she takes hold of my arm and very insistently pulls me back up on deck. And there, just two hundred yards beyond the starboard bow, not one, but two whales are cavorting in the swells. Two huge heads bob fifteen feet above the surface, roll back under, and then bob up again. They are spyhopping, lifting their heads high enough to get a visual cue on what is happening. They seem also to be playing with each other, repeating some kind of synchronized movement that involves a thorough scrutiny of the boat itself. China claps her hands together and decides that these whales are dancing. No, Will counters, they are not dancing; instead, they are making love.

Tom hopes the whales will stay alongside the boat for at least a while longer. He lowers the sails and then rejoins the group sitting on the foredeck. Will drapes himself over the bow to film this event in progress—first the whales bobbing and weaving, then the laughing ladies. Denise has removed her sunglasses. The general mood aboard ship has taken a quantum leap for the better. I begin to howl again in hopes that the whales might hear. All of us humans howl in unison. There is good healthy excitement here. Whales sure know how to make us humans happy. And then, as if in witness to our collective exhilaration, an enormous whale body picks itself out of the water. Thirty-five feet of blubbery muscle suspends itself in the charged air; glistens; and in the bat of an eye rejoins the water with an explosion of spray. It is a signal for us to come together in service.

Seven humans join hands around the perimeter of the foredeck. The urn, like some ancient megalith, rivets our attention into the center. The urn, despite its phony classic Greek patina, conjures up an image of ancient Inca children buried high in the snow of the Andes. Inside large clay pots, tiny bodies mummify in the sparse pure air. After a thousand years, an archeologist comes upon them, and arrogantly breaks the seals as if they are his own property. How does such a violent act affect the souls of the children? Likewise, how is it that Adan is with us today? In ancient Greece, the death of the god Adonis was mourned yearly. Images of him were dressed to resemble a corpse, and then displayed alongside a similar image of his lover Aphrodite, on couches surrounded by flowers and fresh fruit. After a single night, women attired as mourners carried the image down to the seashore, where it was ceremoniously committed to the waves. The Greek mourners did not sorrow without hope, for their songs ensured that their lost one would return to earth again.

During our rite of passage ceremony, the two whales continue to bob up and down, cheek to cheek, like Fred Astaire and Ginger Rogers in *The Gay Divorcée.* Denise's long black hair blows in the wind as she takes hold of the urn, walks over to the gunwale, and offers her dead son's ashes back to the deep. We are all surprised, even shocked—the ashes are not powdery and fine, but rather, bony and substantial like oyster shells. The ashes hang on the surface of the water for a brief instant, then are gone. Each of us sits quietly for another minute, caught in our own thoughts and feelings. But the whales are dancing around the boat. Soon we are up and about watching this impromptu performance.

The performance continues into the next hour. The whales have moved even closer. They have become such an accepted part of the scenery that the two children have begun to take them for granted.

Lounging quietly next to Tom, still clad in my wetsuit, I ask him if it is possible to set sail very quickly and then move right in near the two gray whales. I want to jump in the water. Tom searches my face for some kind of practical motive, smiles, and reaches down to the rope to tie off my ankle. No, I do not want a line this time. I want to swim freely. Tom grins from ear to ear. He extends his hand as if by shaking we are sealing a gentleman's bargain. The sail is set; we move forward. The whales are no longer bobbing. Now they are rolling all over each other, making it nearly impossible to tell where one animal starts and the other finishes. They are kicking up a giant froth. Fifty yards . . . twenty yards . . . I jump off the boat without the drum. Tom veers sharply to the left and away. I am alone in the sea, not quite ten yards away from two gray whales.

## VI. Gray Whales

Since the early 1970s, when tour boats began to visit the Baja lagoons with increased regularity, the whales have begun to change their relationship with humans. In the past, a twenty-yard observation was considered bold. Then by the mid-1970s, the whales began to come right up to the boats and rub their rostrums and backs against the sides and bottoms of the boats. By 1980 the whales had become quite a bit more insistent. It was a common sight to see an individual gray, or even an entire pod of fifteen or more, surround a boat and then one at a time push their head up against the gunwales waiting to be petted and rubbed. A gray whale's skin is largely covered with colonies of barnacles, as well as large pits and scars. It seems a wonder that such a mottled being could even feel the soft caress of a human hand. One well-documented encounter lasted for more than an hour. Sometimes the whales get so insistent that they simply will not let a boat go. And in every single case, the interaction is whale-initiated. No other species of large whale will allow a human being to touch it. In fact, although there are many stories of dolphins that interact with humans, the relationship is actually a very rare occurrence. Last year there were twenty-five such whale-initiated sessions in Scammon's Lagoon alone.

It is not only in Baja that the grays are altering their ways to accommodate a budding relationship with humankind. For years, the gray whale migration was known only by a distant sighting from shore, or a chase by boat. Rarely would a whale linger, or venture, close to shore, except for a brief respite. But now, every year there are new reports of gray whales that choose to linger at various harbors along the coast from Coos Bay, to Bodega, to Tofino. In some cases, it appears as if the whale or whales are house hunting, searching for an alternative to the long Alaska-to-Baja sojourn. This is not so strange when one considers that the gray whales used to live in San Francisco and San Diego bays.

In 1982 a whale stopped near Tofino, along the west coast of Vancouver Island. It too began to rub its rostrum against the sides of friendly boats, awaiting a caress and a pat. This seemed to be the first sign that the Baja cordiality was about to be extended. The whale spent several weeks in the bay. Then one day an ignoramus went out with his dog. He lifted the terrified dog onto the back of the whale as it came close for a pat. The dog sunk its teeth into the back of the whale. The whale dived. The man picked up the dog. The next time the whale surfaced, the man drove his outboard directly over the whale's back, inflicting deep wounds into the flesh of the whale. By the next day the whale had vanished.

Fortunately, for both the whales and us, such sadistic encounters have become as rare as the sight of a whaling ship. Neither species can honestly afford a repetition of the Tofino incident. Gray whales were called "devil fish" by the old-time whalers. They were the only species to consistently attack a long boat when harpooned. In some ways it seems ironic that whale scientists categorize the grays as the most primitive of the whales, a living fossil for its undeveloped baleen and unformed dorsal fin. In many ways their personality seems more akin to the human personality than that of any other whale. They dote on affection, and fight violence with violence. It seems especially critical that, at this period of a seminal human/whale relationship, we take great care to encourage friendship.

For most people, the presence of whales imbues the air with magic. This is no hocus-pocus magic, but rather an interspecies vibration that has proven more than capable of setting the human spirit afire. And certainly, humanity must expend every effort to nurture this connection in order to grow healthy again in an ecologically ravaged world. The word "contact" will continue to come to the fore more and more often. We relearn methods of contact with the environment through the great vehicle of the gray whale. And so the whales spout offshore among their own kind. Along the cliffs and bays of a hundred locales along the coast, the people "oooooh" and "aaaaaah" and likewise feel so very alive among their own kind. From our vantage point high up on the face of a cliff, we laugh together and try to guess where the whales will surface next.

## VII. The Rush

I howl like a wolf, for reasons of safety as much as greeting. I want to be very sure that these whales are aware that there is a tiny stranger in their midst. I swim closer. The surge from a fluke lifts me half out of the water. The coarse, prolific growth of barnacles covers their bodies like a living city built upon a living planet. Are the barnacles aware of the whale? I wonder if these barnacles are consciously arranged by the whales, like some cetacean expression of body makeup? The sentiment need not seem anthropomorphic; after all, the barnacles always seem to flourish in very distinct areas. A phrase of McLuhan floats through my consciousness:

Myth is the instant vision of a complex solution.

I recall the ominous words of a hitchhiker who sternly warned me not to be so foolish as to swim in these shark-infested waters near the Farallones Islands. A diver had been bitten in half just the year before. A fisherman had caught a fourteen-foot, 2,200-pound great white nearby.

Yet I feel no fear while I swim so close to the whales. My confidence may be nothing more than blind romantic idealism, the fantasy of a whale protector with a capital P. On the other hand, I feel there is no reason to be afraid on a day when death transforms into a rite of passage, a great migration. I feel the power of the two whales, the surge of their bodies rolling close like a living wall. Hey, you whales, you got magic? We need plenty! They are ten feet away.

I move in five feet. The whales ostensibly seem to back up five feet. I move in five more feet. Again the whales counter. No matter how fast I swim, these two whales uncannily persist to keep a constant ten feet between us. Their eyes are often peering away from me, so it is not a visual perception. It is totally amazing the degree of control these huge bodies exert in keeping such a precise distance while they roll and wallow in the swells. I had been born with the name Jonah. Was this what happened to him? Was it a gray whale? Why am I so close to these whales? What am I trying to do here anyway? What about sharks? Where is the boat?

Suddenly I was afraid. I needed to get away from these whales, and out of the water as quickly as possible. Attracting whales had been one thing—a noble experiment in interspecies empathy. But this activity of sporting with the whales, it was not right. My goal was a rush of adrenaline and maybe a yarn to tell my grandchildren someday. I had become a thrill seeker. But thrills are a controlled attribute of our fear mechanism. Now I was out of control. The whales deserve better. I deserve better.

This entire inner discourse could not have lasted more than five seconds. In fact, the entire free swim with the whales did not last more than five minutes. As soon as my mood changed, as soon as even the slightest hint of fear came into conscious focus; just as suddenly, the whales dove.

This human being had swum into their space, feeling great excitement. Then he had changed his tune to one of fear. The whales disappeared below the waves. They took a quiet bow to the assembled throng, and then quickly made their exit.

Their sense of timing had been absolutely superb. The boat picked me up. The long, anticlimactic trip back to Sausalito was one of shared songs. Impossible sea yarns, and warm sunshine. Players and audience all agreed that the rite of passage was a rousing success, an event to be remembered. Whales and humans both had learned to comprehend each other's manners a bit more directly.

# Interlude

As I ate lunch, the doe approached, stopped within fifteen feet, and watched as I ate crackers with aromatic cheese and peanut butter.

Her nose in the air, ears slapping back and forth, she stood sideways to me, ready to spring off at the slightest provocation.

I greeted her, "Hi, how are ya," and apologized if I was sitting where she wanted to browse. "I'm sorry that you have to feel so cautious around humans, but I totally sympathize.

"But really, there's nothing to worry about." At this, she changed her posture and faced me head on and watched closely while I snacked on for five minutes. I wondered if my friendly overture might prove her undoing with the next human who walked down the trail. She possessed the most beautiful deep brown eyes.

I broke a cracker and threw it toward her. "Here, maybe you'd like to join me for lunch." As the cracker sailed through the air, she sprang off ten feet, stopped, and then went about her business of slowly easing herself back to it.

She ate it quickly.

Then for the next ten minutes, it seemed as if we just forgot about each other. I sat quietly in the woods with her. That was the best part.

I got up, brushed myself off, repacked the food, and continued my hike up the mountain.

She watched me leave.

# 6

# *Specimen and Participants*

## I.

THE MAKAH INDIANS tell a story about fishermen who are out in their longboats one day when a strange orca with two dorsal fins suddenly appears in their midst. One of the fishermen picks up a ballast stone and hurls it, wounding the animal over the blowhole. The orca swims toward shore, where it is observed beaching itself. But when the fishermen paddle in to examine the animal more closely, instead of an orca they meet a man, who implores them to repair his damaged boat. This they do. The stranger thanks the fishermen, and so launches his boat back out onto the ocean. As the fishermen watch him paddle out, the boat enters a sizable wave and vanishes, to be replaced by two large dorsal fins. The orca blows and then is gone.

This story depicts a belief that is common to American Indian mythology: that animals possess a power to control the way that humans perceive them. The orca had metamorphosed into a man with a broken boat, because that was the only way a fisherman would be able to recognize a being in distress. On another level, this parable about an observer's measure of an event controlling the behavior of the observed sounds suspiciously like an Indian version of Heisenberg's uncertainty principle. And if, as they say, myth reflects life, then this conclusion tends to make hay of the study of animal behavior as it is manifested by contemporary zoologists. As such, the Makah myth may offer some valuable advice to the animal "observers" among us. I offer a personal experience as a means to drive the point home.

## II.

I spent one spring in the jungles of Panama, filming a TV show about communication research between a human musician, namely myself, and highly vocal howler monkeys.

A zoologist, who had been studying the behavior of the monkeys for more than a decade, was asked by the show's producer how she thought the monkeys might react to this music making in their midst. She responded that they would howl back a few times as an expression of territorial prerogative, but nothing more. They would certainly not participate in any improvisatory musical dialogue because they were too dumb (small brain), and too lazy (low-protein diet). She stressed that the monkeys would not climb down from their forest canopy to interact more closely. When humans are about, monkeys never leave the safety of the high branches.

The first morning, I wandered alone out into the jungle until I found a huge ceiba tree filled with dangling monkeys like so much fruit. I sat directly beneath the languishing bowlers and played single notes on a battery-powered electric guitar. No response. Later in the afternoon I tried again. This time at least one monkey responded by urinating from a great height directly on my amplifier.

By the following afternoon I had found myself a broad-brimmed hat, had switched over to the more mellifluous Shakuhachi flute, and tried again. Things began to happen immediately. The entire family howled in close response to the deep resonant notes cast forth by the instrument. Then slowly, the mood shifted. One animal started to fill in the spaces between the staggered notes of the flute while the rest of them listened in silence. One howl and then one note; two notes and so, two howls. This fundamental form of incipient conversation—this dialogue—lasted for about an hour until the approaching darkness forced me to leave.

On the next day the monkeys did not howl along with the flute, neither in the morning nor at dusk. However, very late in the afternoon, the entire family, in precise single file, climbed down from the tree to a branch where they could more closely examine me as I fingered the holes of the flute. There they sat, twelve feet above and in front, watching the music-making process like any rapt audience. Again, the session was ended by the intrusion of darkness.

The next day, the entire film crew traipsed out into the jungle single file, set up their equipment beneath the ceiba tree, and proceeded to film me attracting the family of howlers down from the treetop to a point thirty feet above and in front of my fingers. Some of the monkeys crouched on their haunches. Others hung by their tails with elbows resting on their knees in a bizarre "thinker" pose. There the simian audience remained for five minutes, intently studying the flute playing and the filmmaking. Then, very much at ease, the howlers slowly climbed back into the forest canopy and off into the jungle. They had howled only once.

That evening I described what I had labeled the "dialogue" and the "audience" behaviors to a very intrigued zoologist. No, she had never witnessed anything like it. She concluded that the music must have had a very powerful effect on the monkeys, and that I must have a very special rapport with animals. I agreed with her that the monkeys had certainly responded with an acute curiosity toward the music. But I felt that anyone with basic musical skills, and a benign intent toward the monkeys, would achieve the same results. Beyond that, I explained that any human being out in nature elicits a powerful behavioral response from any animal. Consequently, there can be no such thing as an "objective" scientific observer. Like the fishermen and the orca, we are all participants in an event of

interspecies theater. Humans act out the role of field biologist for the benefit of the animals, and the animals respond as observed specimens watched by human beings. It is nothing more than the uncertainty principle adapted to zoology.

The zoologist vigorously disagreed with my conclusion. After all, such a thesis demeans and destroys the results of all behavioral observation. She then described her own techniques for assuring reliable data. High at the top of her methodological list was the tagging of individual monkeys. I sat up in my chair and may have flinched. To tag an animal you must first shoot it with a tranquilizer dart. The monkey plummets a hundred feet to the forest floor, where it is gathered up into a cage, is carried back to the lab, is poked with needles several times, is branded in some permanent fashion, and is finally deposited back beneath the tree again.

I charged that the monkeys had been so irrefutably observed as staying in the forest canopy over so many years because they were doing their level best to avoid a certain human being who was known to hurt them. Yet they came down from the tree for me, because I had been "read" as being benign. The music cut through a process of introduction that would ordinarily take weeks or even months to develop. The zoologist found my analysis anthropomorphic and thus absurd.

I left Panama the next day, so unfortunately the zoologist and I never got a chance to put our difference of opinion to the test. Would the monkeys have altered their developing relationship with the music if the zoologist had accompanied me to the ceiba tree? Or better still, if she ceased her darting and instead took up the flute, would the monkeys begin to come out of the canopy for her too? On the other hand, would that mean that future "observation" would be deemed invalid by her peers because the monkeys were, somehow, "tame"? Or perhaps the most important question of all: How would the zoologist react if someday she found a man with a broken boat sitting underneath the ceiba tree?

## III.

The relationship between the zoologists, and their discipline of zoology, now stands at a major turning point. On the one hand, we have a school of thought of which our monkey lady is a prime example. This school assumes, a priori, that Mankind resides uniquely alone and separate from the other animals—at the pinnacle of evolution. (Animals may not be machines, not exactly machines; but then, neither are they emoting peers of human beings either.) The members of this school also believe that they can somehow stand apart from the environment, and watch life unfold from the invisible perch of their objective point of view. This school views the animals, and in a way, the entire world of nature, as *Specimen.* It is an important term, a powerful word of separation; and it sums up the relationship between the members of this Specimen School, and those creatures they choose to observe. Specimen are to be collected, experimented upon, killed, mounted, and discarded, all in the name of the pursuit of knowledge. And what God was to the scientist-priests of the Middle Ages, the pursuit of knowledge is to the Specimen School of zoologists of our own day. And "pursuit," of course, is, first and foremost, a word of aggression.

While this pursuit of knowledge syndrome may have given our civilization much information, much fact, about the mechanics of all the various *components* of nature, it did little or nothing to further our understanding of the whole. In fact, it was never made totally clear if we humans were, or were not, parts of that whole. After all, we were above, and thus more than a bit separated from, the rest of nature. Practically speaking, how could a zoologist observe even an inkling of the Big Picture when the majority of the research was being conducted in the laboratory on dead and caged animals? The published observations gave us all a criminally flawed view of the natural world.

It is interesting that, although post-Darwinian zoology has quite thoroughly refuted the Biblical scheme known as creationism, it still borrows much of its relationship with the animals from the book of Genesis. That is, animals have been placed upon this earth for man's use and enjoyment. This easily translates as the need, yes even the right, to experiment upon animals to further the pursuit of knowledge. And until very recently, this long-prevailing attitude effectively overrode any other human/animal ethic that attempted to establish a different, more benign relationship with the animals. The Specimen School had amassed great powers in our culture. The lawmakers turned to them as the stewards and spokesmen for national and international policies toward the environment. And so, most of our legal, political, and even cultural views about animals unfolded from this base of animals as Specimen. Making the Specimen zoologists our stewards of nature was one of the most grievous wrongs that we, as a culture, ever perpetrated upon the environment.

Just as the Industrial Revolution spawned a powerful system of "expertism," so the members of the Specimen School were, and in some cases still are, our animal experts. The depiction of animals as Specimen, tethered to the philosophy that the pursuit of knowledge justified any activity, fit itself perfectly to the needs of an emerging civilization hungrily disposed to tap into the seemingly bottomless wealth of nature. Just as the zoologists had their specimen, so their brothers in business viewed nature as one marvelous *resource.* And just as science was guided by the pursuit of knowledge, so the exploiters believed in their noble vision of *progress.* Yet, as we all now know so well, something soon went awry. The exploitation of nature, once considered a benchmark of progress—that is, leading us forward—can now be understood only as leading us to the very brink of destruction.

But by now, it has become a truism to blame all the excesses of Western, exploiting man on that misguided, poorly understood principle known as progress. What may not be so well known is that the very idea of progress—as an upwardly mobile improvement over time—is decidedly Western, and more so, quite modern in its interpretation. Many cultures believe that progress is a winding down. Seyyed Nasr has commented:

> *Traditional* Western Man, like his fellow human beings in various Eastern civilizations, saw the flow of time downward, rather than in an upward direction.

And thus, we Westerners believe that our machine culture *should* be getting better because we ourselves are bound by the cultural bias of the industrial revolution. Yet by the standard of much Eastern thought, our accelerating use of technology—ever more destructive—is a course right on target. If there is any truth to be found in any of this, it is that we need to slow down the pace and realign ourselves with a more direct relationship with nature. Disaster is our alternative.

If the Specimen School developed hand in hand with the needs of the nineteenth-century industrialists, then there is also a much older school of science, whose attitudes are just now beginning to be reinterpreted. This is the way of the shaman. Joan Halifax, in her excellent book *Shaman, The Wounded Healer,* defines the traditional creed this way:

> Although the universe is conceived of as powerful and uncertain, it is also a cosmos that is personalized. Rocks, plants, trees, bodies of water, two-legged and four-legged animals, as well as those who swim or crawl—all are animate, all have personal identities.

The shaman incorporated the priest, the doctor, the psychiatrist, and the scientific generalist, all in the persona of a single tribal member of great authority. Just as the Specimen School had developed to fit the moods and bias of the industrial revolution, so the shaman's science fit the moods and bias of preliterate and preindustrial culture. The Specimen School broke nature down into its reduced components; the shaman dealt with a whole, in which humanity was but one ingredient. The Specimen School took license to exploit the natural world to satisfy an insatiable hunger for information: knowledge as power. The shaman already possessed that power because he or she was guided by a worldview in which all animals had rights, dignity, and enormous powers of their own to manipulate the destiny of us mere humans. The shaman alone knew how to deal with them all as equals. More so, the power of the shaman was shared within the culture as a form of *medicine,* used to keep the entire tribe in balance with the forces of the environment. This balanced relationship between power and healing was key, and also very straightforward. No one thing was healed apart from everything else being healed at the same time. After all, humans and animals were viewed as different manifestations of the same spirit. This, from the shamanic culture of the Nalungiaq Eskimo:

> In the very earliest times, when people and animals lived on the Earth, a person could become an animal if he wanted to and an animal could

> become a human being. Sometimes they were people and sometimes animals and there was no difference. All spoke the same language. That was the time when words were like magic. The human mind had mysterious powers. A word spoken by chance might have strange consequences. It would suddenly come alive and what people wanted to happen could happen—all you had to do was say it. Nobody could explain this: that's the way it was.

If this description of the environment of a shamanic culture seems more appropriate as the theme for a fairy tale than as a zoological postulate, then compare it closely with the statements made by the monkey zoologist in 1983. Which assumptions are truer? In fact, both attitudes express a truth only as they relate to the cultural context in which these attitudes are based. Do howler monkeys always live high in the trees? Well, for some zoologists they certainly do. Do people and animals speak the same language? The Nalungiaq Eskimos certainly believe that they did long ago. Or for that matter, do howler monkeys come down from the trees and *also* speak the same language as human beings? They evidently do just that for some people who know how to converse in the language of music.

Consequently, no version of each of these individually cherished beliefs can effectively explain any kind of "ultimate truth." Instead, information itself must be held suspect. And of course, we have now come full circle back to the point where we began this discussion. The natural sciences, because of any number of semantic, historical, and cultural biases, can never provide a direct window to the truth of nature. If it persists in representing itself as truth, then it is a fleeting kind of truth—a cultural truth—a truth fit for the times in which they are recognized as such.

After the atom bomb, after Earth Day, after it became self-evident to many people that our contemporary relationship with the environment was amiss, a new breed of zoologists began to emerge to work with the new situation. A new mood took over the science: a strange brew of the general methodology of the Specimen School wed to the very internalized human/animal unity of the shamans. The myth of the impartial, invisible observer of natural phenomena was slowly being replaced by the creed of the participant. We humans were an active part of the natural process after all. Furthermore, because this new *Participants School* had emerged in response to the debilitated environment, the bulk of the new research was conducted only with wild animals out in the field. The attitude that had for so long depicted animals as specimen was now being replaced by a new methodology known as benign research. This meant, in effect, that no harm was to be done to the animal subject, its society, or its environment. Benign research was *for* animals. The Participants School of zoologists were doing their level best to help the animals to survive and rehabilitate in a world turned upside down by the excesses of human exploitation. Like the shamans long before them, the Participants usually view themselves as *healers.* It is a positive, life-affirming attitude, a position that will, no doubt, continue to win converts. And there is so much healing to be done.

Just as there is, at present, a constant and sometimes escalating struggle going on between exploiters and environmentalists, so, similarly, is there an ongoing, if not quieter, struggle going on between the two schools of zoology. It is a natural outcome of what Thomas Kuhn has called a paradigm shift. How will it finally solve itself? Physicist Max Planck, surveying his own career as a scientific innovator, expressed it this way:

> A new scientific truth does not triumph by convincing its opponents and making them see the light, but rather because its opponents eventually die, and a new generation grows up that is familiar with it.

In my own field of cetacean research, there is almost no study anywhere that does not display at least some signs of the changing of the scientific guard. The reasons are obvious. The cetacean scientists of the Specimen School were sponsored in their research primarily by the whaling industry, the military, and the oceanariums.

In the first case, the whales were slaughtered to the very brink of extinction, all the killing justified by the "expert" opinion that no lasting harm was being done. It was the whaling establishment who gathered together all the statistical data about whale size, weight, anatomy, and physiology. Yet, ironically enough, because all these studies were done on dead animals, no accurate picture of what a living large whale looked like was made until the first of the Participant zoologists actually jumped into the water and looked at them. The image of the balloon-shaped whale does not exist anywhere in nature except in the flesh of a dead, distended carcass pulled up onto the slipway of a factory ship. Even when a baleen whale feeds, it looks more like a giant tadpole than a balloon.

In the second case, several species of large dolphins were trained to retrieve various metal objects, even torpedoes. It was the zoologists employed by the military who first gave us an inkling, through dissection, about the large brain of the dolphins. The military zoologists were the first ones to explore that unerring ability of dolphins to be trained in clever ways, undreamed of with other animals. And it was from the ranks of these torpedo-retrieving, and possibly torpedo-detonating, animals that we first learned how a dolphin might commit suicide if it became too depressed by unabated captivity and separation from others of its kind.

In the third case, the bulk of the early behavioral studies, including interspecies communication research, has been accomplished with dolphins affiliated with oceanariums. These studies gave us our first glimpse of the complex social behavior and probable language of the dolphins. It was at the oceanariums that the human lay public was first able to see for themselves just how incredibly playful and astute a dolphin could be. In fact, many oceanariums justified their very existence in terms of the educational service they provided for the public. But ultimately, every single oceanarium performance or research study also demonstrated how a normal, vibrant, complex, social animal behaved when kept captive in a tiny, concrete enclosure—forced to perform stunts for food. The "truth" of the oceanarium dolphin was a vision of a paid entertainer, whose main purpose in life was to perpetuate the ruse of the happy specimen.

Perhaps the picture that I paint here is too grim. It is thus essential to point out that once the changing of the guard began in earnest in zoological circles, it was more an evolution of cultural priorities than of the discipline suddenly metamorphosing. The Specimen School provided the background information, the fuel, for the fire of the Participants. And in all fairness, it should be strongly noted here that these two schools are not tied down to any clear-cut era. This is not exactly old vs. new. There were Participant zoologists out in the field a hundred years ago. Likewise, there is still a trickle of Specimen biologists being turned out by our universities today.

And with that in mind, I have left the most controversial example of the Specimen School for last. It is easily the subtlest, and therefore the most insidious. It may be the most self-defeating. It answers the question of how the Specimen zoologists work to save animals from extinction. How they heal.

There is a popular American television series about animals that weekly demonstrates attempts by scientists to save various endangered species. One week the host is seen in the Canadian Arctic with the lynx or the wolf; a week later he is observing a zoologist who works to protect the last of the Madagascaran lemurs. The general mood is the same, week after week. We watch a zoologist out on the African savanna with a notebook in one hand and a tranquilizer gun in the other. He is a scientist in the employ of the developing country who owns this savanna. Bang! A cheetah is darted. She soon sways and falls into the grass in a drugged swoon. Blood samples are taken, perhaps a liver biopsy, possibly a tooth extracted to determine age. Then the animal is securely trussed, caged, and driven overland a thousand miles to a preserve. The preserve is securely fenced and guarded, offering full protection from the excesses of overpopulating natives as well as poachers. These are the twin destroying angels of the cheetah race.

Thus a few lucky cheetahs are saved. The government has wanted to open up that old cheetah savanna to agricultural development. The original habitat can now be overrun by farmers who have already begun to do it anyway. The zoologists, as members of the scientific community, give the entire operation an air of professional credence.

But if this scenario is healing, then it is the same kind of healing as putting eye drops into very red eyes without taking any measure whatsoever to cure the causative allergy. The eyes look great for a few hours, but the disease is still present in the body. And in both cases, whether we discuss allergies or cheetahs, the Specimen School relies on the treatment of *symptoms*, and throws their collective hands up in the air when the subject of true prevention is brought into the discussion. After all, at least as it applies in the relationship between cheetahs, developing countries, and farmers, the Specimen zoologists have no license to control the movements of overpopulating humans. Only cheetahs. Furthermore, the Specimen zoologist believes, a priori, that the humans have the prerogative to move wherever they please. And of course, it may be true that if he spoke out against opening the savanna, he might simply be replaced by another scientist willing to accede to the government's wishes. Thus government and science bounce off one another to achieve whatever job they set out to do.

Certainly, it is a very hot issue, not one solved without a comprehensive course of prevention. It demands a person, a group, or an entire movement willing to take a stand to heal the environment through participation and prevention.

The Specimen School has its methods, attitudes, and its overall mood; so the Participants School has its methods, attitudes, and moods. There are many examples of this new paradigm in action.

During the mid-1960s, a whale scientist, Dr. Roger Payne, conducted one of the very first benign research projects with large whales ever undertaken in the wild. Payne discovered that humpback whales possess a highly complex vocalization behavior that seems more reminiscent of human songs than of stereotypical animal calls. Over several years of observation he learned that the humpbacks off Bermuda possess a markedly different song structure than the humpbacks off Hawaii. And in either location, individual whales might sing for fifteen minutes or more, and then repeat the entire song over again. This year's song displays close similarities to last year's version, with slight but clearly discernible differences. Here was a very clear instance of an oral tradition. It implies that the humpbacks possess at least the rudiments of learned culture.

This single discovery did more to mobilize the efforts to save the whales from extinction than any other research study of its time. The whales could sing! Payne himself, obviously deeply affected by his discovery, became a major spokesman in the fast-developing save-the-whales movement. And what, we all wanted to know, were they singing *about*?

Payne's study, as an exercise in scientific method, was actually quite orthodox. He sought information, and found it by gathering a large-enough database from which to extrapolate results. But his *attitude* was very different from that of almost anyone else who was working with whales at the time. For one thing, he focused entirely on wild, live, socially interacting whales. To some observers, the humpback study seemed more in the spirit of an anthropologist studying a human culture than anything that might be construed as Specimen research. Payne sought whale preservation, and thus, utilizing the techniques of benign research, he made every effort not to disturb his subjects. There was also a practical consideration to this. After all, they might have stopped singing had he been too intrusive.

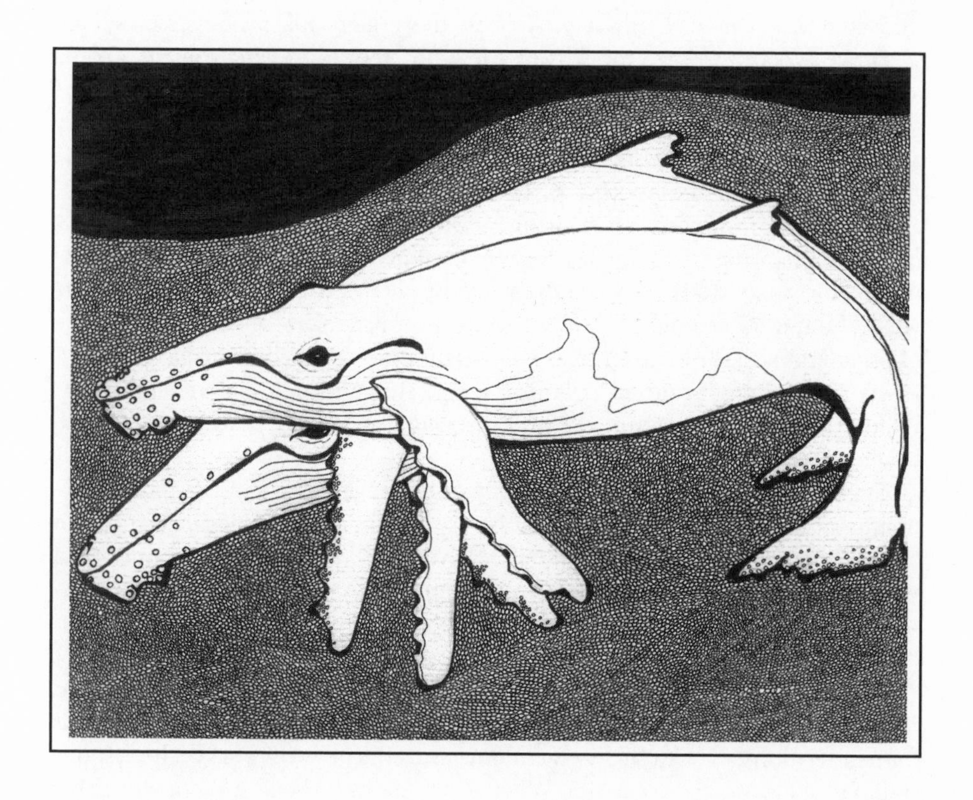

The results of this study sharply demonstrated that the old Specimen version of "whaleness" was not only dim—the product of dead animals and an exploitive modus operandi—but incredibly trivial as well. Payne's new "whaleness" gave the rest of us landlocked humans our first glimpse (whisper) of a living, vibrant being. And it seems no coincidence that Payne released his database as a record album, at a time, in the 1960s, when our human musicians (Dylan, the Beatles, etc.) were very definitely pacesetting the cultural mood of the moment. Thus the information became plainly accessible to those laypeople who possessed the incentive to work to save the whales. How different a presentation than some article presented in a professional journal, to be scanned only by the very few professionals doing similar cetacean research. In conclusion, Payne's classic study shares as much with the healing attitudes of the shamans as it does with the methodology of all of modern science.

The vision of the whales had changed, and with it the mood of the times very quickly transformed from apathy to mobilized preservation. One can only wonder: If this single study could so dramatically change our cultural image of the whale, how, in turn, might any other research project likewise change the image of other animals? Or to generalize, how might they change our image of *nature*? Lewis Thomas has achieved a similar image shift through his writings about microorganisms. Who among us will instigate a new study and image for the belabored cheetah?

Thus, because science is as much a creator of image as it is a purveyor of information, it serves culture in a way that is very similar to art. Copernicus initiated a paradigm shift by showing that the earth revolved around the sun rather than vice versa. Similarly, Payne initiated a new image of the humpback whale. Likewise, Bach and Picasso created new images of sound, pace, and rhythm; or color, light, and shape.

And although zoology may not put us as close to the "unadulterated truth" as many zoologists still wish to believe that it does, it certainly serves to define our human relationship with the ways and means of the physical world. The responsibilities of a zoologist are major. Whether we are to move quickly or sluggishly from our human role of blind exploiter to a role of steward is very much in the hands of the information and *the mood* provided us by our zoologists.

# 7

# Behold, the Ring of Light

*Reason, I know, is only a drug, and as such its effects are never permanent. But like the juice of the poppy, it often gives a temporary relief.*

—William Morris, *The Wood Beyond the World*

## I.

THERE ARE LITERALLY trillions of possible pathways available to the human nervous system. As the fetus transforms into a functioning human being, these connections begin to fire in response to all the history and chemistry carried within our genes. When we emerge from the womb, the connections continue to route themselves into roads and well-worn byways as a result of all our varied influences, whether they be education or accident, mundane routine or fantastic mystical vision. Over time each of us accumulates our individualized likes and dislikes, our talents and our problems. The world's two premier chess players square off for a championship match, and utilize entirely different sets of connections to make the very same move.

It is a common fact that human beings use but little of the enormous potential capacity within the brain. We systematically develop certain key areas, while other areas lie fallow like some vast unexplored wilderness. Likewise, some people have developed their brains in ways that others have not, or cannot. In certain cases these individualized connections involve modes of consciousness that we generally lump together under the heading of extrasensory perception. Our culture chooses to label these perceptions as extrasensory because their effect is impossible to correlate with our normal sensory apparatus.

Continuing our geographic analogy, ESP may result when a certain inimitable combination of brain waves veers off the main highway. We find ourselves riding along a bumpy and little-traveled path that passes through some magnificent countryside. The next time the wave current enters that part of the brain, it veers off again, preferring the view to the speed of the interstate.

ESP is one of those subjects that is easier to accept than to explain. The explainers within our culture are the scientists among us. What they cannot properly measure, nor succinctly describe, they also find impossible to acknowledge. Thus,

ESP exists as a subject of heated controversy. In a way, there is a kink built into the system. The Cherokee medicine man, Rolling Thunder, describes the problem this way:

> The rational mind is a cataloguing and comparing mechanism and it sees only what it contains.

There lies the crux of the reason why ESP is so difficult to either prove or disprove. It resides in some intellectual limbo—an intuitive no-man's-land halfway between the rational mind and mystical vision. That is why the skeptics often seem to become mired in their heavy-headed verbiage, why the mystics end up rolling their eyes and keeping silent.

If humans do possess these varied abilities, this ESP, does it seem any wonder that the animals possess it as well? And of course, if it sometimes seems difficult for us to accept these special abilities in our fellow humans, then how much more difficult still to accept them in other species?

If we take the point of view that there is a deep intellect and communication network present throughout nature, then there must be signs to read everywhere. In some cases, these signs emerge as little more than faint whispers spoken in a very foreign language. In other cases, we look directly at the tip of the iceberg, fully aware that the major portion of its mass lies beyond our grasp to explore. Sometimes we may become fleetingly aware of an extrasensory perception in another species, but fail to recognize it as such.

An example of this is the ability of birds to navigate by sensing the earth's magnetic field. Another case might be the great white shark's sensitivity to electrical fields. The first white shark ever kept in captivity, at the San Francisco Aquarium, kept bumping into one spot on the side of the pool. It was discovered that there were cables buried into the wall at just that spot. The shark became so confused at judging distance, and continued to bruise itself so incessantly, that finally its keepers decided to set it free.

It is easy to argue that neither of these examples has anything extrasensory about them. Rather, they are the quite normal perceptions of that particular species, rather more analogous to our human sense of taste than to, say, psychokinesis. But it is not my meaning here to argue whether or not an electromagnetic sense does fit into the accepted niche of ESP phenomena. Rather, I take the point of view that all animal perception, including human perception, sensory or extrasensory, is the result of normal species development. We humans possess the potential for bio-navigation; we have simply not yet developed those connections in our brain. We name our evolving powers "extrasensory" only because our science has not quite figured out how we actually achieve some of the uncanny effects. It is not that clairvoyance defies the laws of physics, only that our evolving physics has not yet gathered enough information to put forth an explanation. We search for a tidy explanation of a bird's navigation abilities, and so we find it. But perhaps that is only part of the picture. Perhaps the birds also tap into the universal minds of all their ancestors who made the same trip. Farfetched? Possibly.

When we make our way forward, keeping an experiential, positive, and participatory attitude as our basic working method, nature begins to open up more fully before us. She begins to share her behavior, her talents, and yes, her mysteries, in fresh and unsuspected ways. A few examples may suffice to bring this point home.

We were sitting on the front porch of Old Man Nicholas's bamboo and thatch house. Nicholas was half Indian, half something else; in his mid-sixties; the proverbial trickster, Castaneda's wise man; sometimes spouting dirty old man prejudices, other times dazzling us with firsthand anecdotes about unknown worlds. It was dusk; the coconut palms out on the cliff were flapping briskly in reply to the sharp offshore breeze. But where we sat, strumming on the banjo and watching the sun set over the Pacific Ocean, there was no wind. Nicholas was in especially fine form that night. He had said: "You think the trees are green? No, they are only sometimes green. At night they are black."

All of us on the porch knew that this evening's fiesta was just about over. Within fifteen minutes we would be forced to retreat to the protection of our own well-screened palapas. There was simply no other escape from the dense hordes of mosquitoes who also called this seaside jungle their home. Nicholas himself was only too aware of the dilemma. He was always the consummate host, an inveterate celebrator who had seen just about every party he'd ever given succumb to the unrelenting invasion of these tiny and very swift insects. Eventually, all the rest of us—male, female, Mexican, and American—would be forced out. Nicholas would spend the rest of the evening alone on his porch, whittling and whistling. For some reason, the man never got bitten by mosquitoes.

The next morning was glorious with warm sunshine. I sat on the cliff all alone, studying the long tubular waves that had made this beach a haven for American surfers. After a while, Nicholas shuffled over in his characteristic way, and sat down on his heels beside me. I asked the old man how he escaped the mosquitoes. He chuckled in that unique way of his that signaled to me that he thought it was funny the way gringos made magic out of something so totally obvious. He held out his arm in front of my face and proceeded to clench his fist three times. Then he asked me to feel his forearm muscle as he clenched again. "It's a little bit like that," he began. "A little bit, but not exactly like that. You make the blood flow through your body in a certain controlled way. The mosquitoes, they know the language of the blood better than anyone else. I've spent forty years here, so I've had a lot of time to learn that language too. Now, I know how to tell the mosquitoes to stay away."

Years later, while reading Doug Boyd's *Rolling Thunder*, I came across another novel method for keeping mosquitoes at bay. It accomplished the same end, but

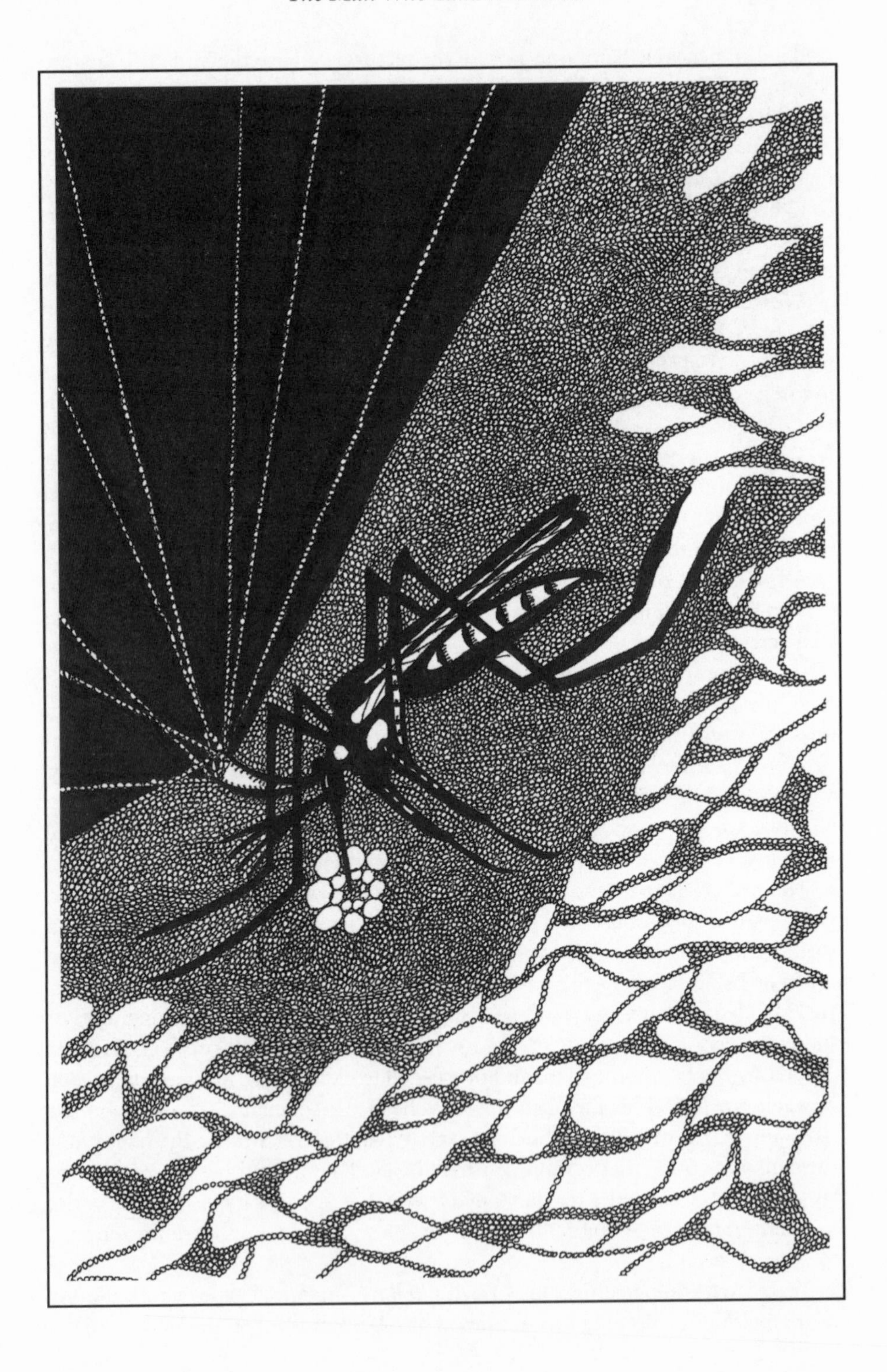

employed a totally different kind of extrasensory perception. Nicholas and Rolling Thunder were like the two chess players who invented the same move from totally different parts of their psyche:

> Mosquitoes won't bother you—they might not even touch you—if you know how to maintain your good feelings. These attitudes make vibrations; and they have a smell to them. That's what keeps the mosquitoes away. . . . It's not so easy, that kind of control. But it's not impossible because you do it yourself. It's all done from the inside.

Nicholas, on the one hand, learned to communicate directly to the mosquitoes utilizing a kind of pulmonary Morse code. The ESP involved had developed from a precise, parasympathetic body control wedded to a quite remarkable display of interspecies communication. Nicholas drummed out a threat and the mosquitoes responded. After I finally left Nicholas's seaside enclave, I often wondered if the old man could have just as easily cajoled the mosquitoes to bite *only* him. It would have been a valid proof of his method. Yet although I did acknowledge Nicholas's talent to speak to the mosquitoes, I somehow doubted that he could actually reverse its effect.

Rolling Thunder, on the other hand, took a more benign and trusting approach. The emotions that had produced the good feelings also produced a repellent that kept the mosquitoes away. Insecticide concocted as crystallized intuition. As a relevant footnote to this matter, it seems no accident that both Nicholas and Rolling Thunder believed that all animals communicate with one another through telepathy.

Yet interview some of the world's most inspiring musicians and you will no doubt discover that the communication between them, while playing, involves a great deal of telepathy. This is such an accepted attribute of good ensemble playing that the musicians themselves, more often than not, treat it all quite matter-of-factly. In my role as a guitar player in a rock and roll band, I sometimes wondered if the telepathy was simply a hidden vehicle for good music making, or whether the music was, in fact, an acoustic rendering of the telepathic connection. Music is not only the food of love; it is also the current hit from our own bio-radios.

It is the same when one plays music with an animal. The act of creating harmony cuts deftly through all the obvious and monumental differences between the species. Long ago, working with that tom turkey in Mexico, I somehow found myself entering a realm of mutual expression with the bird that far transcended the recorded result of a few well-placed gobbles. I became the turkey and the turkey became me. It set off fireworks in my consciousness. Yet what it was that the two of us actually communicated is much more difficult to pinpoint. It was not words. Nor was it emotions either. Not exactly emotions. Rather, it was that intuitive limbo again: the exchange of pure energy, the kind we generate so effortlessly when we are young and in love, it is the stuff of talented musicians at play. For me, the experience was immediate and very direct. It shattered all my preconceptions about turkeys, about animals, and about my hidden ability to transmit and receive telepathically.

In my human-chauvinist manner, it took years before I could actually say to another human being that I had received the seminal, mystical experience of my life while sitting in a barnyard, gobbling with a tom turkey.

## II.

> *"You can see this wall if you look carefully. It is a dark line like smoke. And it hangs there between [the fighting cocks] until something pushes it away. Then the cocks hit each other with their feet. And this one over which the smoke passes will always lose. We all know this."*
>
> *I didn't. I felt naked and did not know what to say, but she wasn't expecting a response.*
>
> *"Dogs and buffalo do the same. And people. Whenever you grow angry in the school, your smoke covers the whole class. . . . Fear and love and hate all make shadows, but their colors are different. And all blow away as smoke will in the wind unless they touch another of their color and meet to form a wall."*
>
> —Lyall Watson, *Gifts of Unknown Things*

I went to Yellowstone National Park to study and film the annual bugling ritual of the elk. Bull elk really do bugle, and it is a sound that, if it did not actually occur, could never be invented. The Tarzan-like call is easily one of the most haunting sounds in all of nature.

A bull elk begins bugling (not honking, not trumpeting, but bugling), and all the females within earshot come running to him in sexual anticipation. However, if another bull in the vicinity also begins to bugle, then all the females will leave the side of the first bugler and run over to join the second singer. Of course, the first elk makes a grand effort to stop all the ladies from exiting so fast. He tries to stall them by frantically throwing his body between them and the new challenger.

Eventually, as the morning wears on, a harem of ten or more females may come together, all dashing here and there between two or three competing bulls every single time one or another of them lets loose with a romantic bugle. Naturally, this process of bugling, rushing, and blocking tends to bring the action inward. As the bulls move closer and closer to one another, they get their first close-up look and smell of the competition. One look at a better-endowed competitor may be quite enough to prompt a smaller bull to call off his own performance.

Eventually, there may be two like-sized, like-minded bulls left in the bugling competition. They are now very near to one another; the females have begun to

slow down from the exhaustion of bouncing back and forth between the two bulls. The bulls stand facing each other with a group of females milling around between them. Now is the time to begin the famous antler jousting of the bull elk. However, the bulls do not start at a distance and run full tilt at one another in a loud smack of skull and antlers. Rather, they meet one another, lower their heads, and push forward, more like sumo wrestlers than medieval duelers. Rarely does a bull draw much blood. Far more dangerous is the potential threat of locking antlers. When that happens, unless the two bulls work together to extricate themselves from their Chinese puzzle of antlers, both will die of starvation.

But much more often, the most ebullient bugler will humiliate his lesser adversary long before they get to the point of body contact. This is more a battle of the bands than a Western-style showdown on Main Street. The winner lets out one last triumphant bugle, and then trots off with ten or more wives by his side for the rest of the day. So begins the process of the perpetuation of the elk species. And tomorrow morning, the bugling once again begins in earnest. Today's exhausted lover may very well become tomorrow's vanquished bachelor.

This ritual occurs only during the few weeks in the fall when the temperature first dips below freezing. The temperature drop seems at least partially responsible for stimulating the hormone that controls the heightened sexuality, the bugling prowess, and the very heavy, crazed breathing that accompanies them. Here is a masterstroke of sexual timing—rutting during this brief period in the fall ensures that the babies are born at the proper time next spring.

An American television show had sent me to Yellowstone to attempt to interact with the elk. Essentially, I was to try to woo and win some female elk through the noble art of bugling. It was a losing proposition. Every morning I woke up long before dawn and commenced to jog out onto the vast open woodland of the park. There, I would soon spy many elk, both males and females, roaming about in the subfreezing splendor of dawn. Soon, I would hear a bull bugling from a half mile or more over the next rise. I'd sprint off in his direction, stopping only after I could watch him from the concealment of a large tree trunk. He'd bugle again. Then I'd wait for thirty seconds or more until I was absolutely sure that there were indeed females running to join him. Just as they arrived by his side I would let out a loud shrill blast on my tenor recorder: C-E-G-C, from low to high. The bull would turn, and the females would start running in my direction. But they would get no more than halfway to my hidden position; look around for the welcome sight of a proud, panting, sweaty bull; see nothing at all; and soon head back to the real elk. I could tell that my imitation bugle was, in fact, quite convincing, because the second time I let fly, the females came again. This time, I stepped out from behind the tree. I might as well have been Godzilla descending over Tokyo. Seven females and one bull elk took off so fast in the other direction that I did not hear another bugle for more than two hours. Such is the reality of humans who enter the woods carrying long, tubular objects.

But I felt bold. I believed that one solitary bugle, causing one headlong rush by hot female elk, if set up properly, could give us the film sequence we needed for the show. My verbal explanation of the entire courtship ritual would serve as a strong filler to keep the images comprehensible and educational for the viewing audience of 20 million or more people worldwide. And after all, on my second morning of experimentation, the females had actually come within twenty feet of me before making their getaway. I found a telephone and called the film crew. They arrived twenty-four hours later.

The morning they arrived from Los Angeles heralded the beginning of Indian summer. The weather was magnificent—sunny, clear, and oh so warm for this time of year. Unfortunately, the warm weather was a harbinger of disaster for my hoped-for career in network television. Over the next three days the temperature never once dropped below freezing. There were no antler jousts, no crazed panting, no bugling; there was no show.

However, there was still a ray of hope. After all, Yellowstone contains more varieties and higher concentrations of animals than probably anywhere else in the United States. During my four full days in the park I'd already seen more kinds of animals than ever before in my life. There was the silent, complacent porcupine munching on a tree branch while sitting comfortably right in the middle of a dirt road; a silent herd of bighorn sheep grazing right next to the main road just below a high mountain pass; several gigantic moose letting out an occasional bellow as they bounded off noisily through the underbrush; trumpeter swans trumpeting on the opposite shore of a steaming lake. And yes, the swans did trumpet in response to some energetic harmonica playing, but the aesthetic correlation between blues harmonica and elegant swan trumpeting was at best obscure, and never anything to promote on national TV.

However, during my third afternoon in the park, while driving through an immense open prairie surrounded by high snowcapped mountains, I chanced upon a herd of at least a hundred buffalo grazing several miles from the road, beside a wide but shallow river. It seemed a promising image for the show, and an intriguing study in interspecies communication. Even if the buffalo made no vocal response to human music, they might react or relate in other ways. I couldn't know for sure until I tried.

Before the arrival of the white man on the American continent, there were as many as 60 million buffalo grazing throughout the United States from New York to Georgia, and then all the way out to the West Coast. It was probably the highest concentration of large mammals found anywhere on the earth. There were two distinct subspecies: the plains buffalo, and the larger and darker wood buffalo. A large bull wood buffalo may be twelve feet long, stand six feet at the shoulder, and weigh twenty-four hundred pounds.

The last buffalo east of the Appalachian Mountains was killed in the state of Pennsylvania in 1801. Yet in the year 1830, there were still as many as 40 million

of the animals roaming the Western plains. Then the killing began in earnest. At the beginning of the slaughter, buffalo were killed for their meat, and for their hides.

Then something went very foul; the most heinous mass extermination of any large animal species, before or since. The buffalo were killed simply to clear them off the face of the plains to make room for the new white settlers. They were also killed as a means to starve out the several Indian tribes who were wholly dependent on the buffalo to provide food, clothing, and even skins for shelter. Massacring the animal became a kind of nineteenth-century form of outdoor recreation. Politicians opposed any efforts made to preserve the great herds since killing them was generally viewed as a patriotic thing to do. Buffalo Bill Cody became a national celebrity for his exploits as a buffalo hunter. He boasted of killing 4,280 of the animals in one year. Once in a while he would eat the tongue of a buffalo, which was considered a great delicacy. Usually the carcasses were left to rot or feed the wolves, who were also on the national hit list. It got worse. In 1872 the plains turned a deep blood red. More than a million buffalo were killed per year for the next three years. By 1889, there were only 541 buffalo left, most of them living on the high prairie of Yellowstone Park. In 1891 there were 300. In the year 1900 there were only 39 buffalo. Thomas Berger, in his book about the Sioux, *Little Big Man*, gives a moving account of how the Indians viewed the white hunters:

> The human beings [the Sioux] believe that everything is alive not only the men and the animals, but also water and earth and stones and also the dead things from them . . . but white men believe that everything is dead: stones, earth, animals, and people, even their own people; and in spite of that, things persist in trying to live; white men will rub them out.

But almost miraculously, the buffalo recovered. In 1900 stern measures were taken to manage the preservation of the last small herd. Today, over thirty thousand of them exist throughout North America. The preservation of the buffalo is generally considered to be one of the finest victories for wildlife preservation on the American continent. Today, there are about eight hundred buffalo roaming the high open plains of Yellowstone National Park.

The cameraman in the film crew, a longtime Wyoming wilderness professional, shuddered when I explained my premise of wandering down into the midst of the buffalo herd. "Buffalo are dangerous animals. If you get too close to the herd, the alpha bull will charge; and out here on the prairie there is no place to run to." I had not known that. More so, I believed that his point of view could be true, but like other animal hearsay such as the taboo against swimming with killer whales, it depended entirely upon one's own relationship with the animals. Buffalo might certainly flee from a hunter, and charge a cameraman; but might they not react differently to someone who projected both music and friendship? I believed that the charge of a bull buffalo was the mindful act of a herd leader trying to protect his herd from a source of potential violence. Where there was no threat, there was no need to protect.

I felt that the key to my success depended upon whether or not I could flow into the same time frame as the buffalo herd. That is, they seemed slow and unhurried. If I met them just as unhurried, might they not accept me? I was bold—I was sure that the buffalo would accept me. The director of the television show agreed with me, no doubt convinced by the amount of money he'd end up losing without a finished film. However, the cameraman rightly insisted that our finished segment very plainly state: "Not everyone should try this."

To woo the buffalo herd into accepting my presence, I chose to play a simple repetitive drone on my onion-shaped guitar known as a vihuela: A-E-A-C played over and over again without any discernible rhythm. Next, I asked the film crew to set up their cameras and tape recorders on a small knoll four hundred yards downwind of the herd. It seemed critical to the success of our venture that they shoot the entire sequence without moving and maneuvering for shots. When everyone was set up, the cameras with their long telephoto lenses, I jumped over the stones and fossils that littered the shallow river bottom, began playing the drone, and slowly moved in on the herd.

It took well over an hour to advance the first three hundred yards. I tended to move when the buffalo moved, and the entire herd would spend long minutes frozen to one spot, chewing and chewing and chewing. Then one animal would move ten steps to the right or left, and ten others would follow. I began to dance in place, a reaction to something that I had once read about the poor eyesight of the buffalo. If they knew that you were there and could follow your movements, they would not become suddenly alarmed. It was only after they *first* saw you, after you were already close, that they reacted. But by dancing, moving my arms and legs very slowly in time to the music, I left little doubt that, indeed, there was somebody advancing toward them.

Slowly, ever so slowly, I shortened the distance between us. Then something strange happened. Suddenly, almost too suddenly, I stood no more than a hundred feet from the nearest animal. It was as if, all at once, I had been lifted right into place, without any recollection whatsoever of the long arduous hour of moving forward at a snail's pace. Perhaps it was nothing more than my sensory reaction to the very hypnotic drone. Perhaps it was the shock of staring directly into the big brown eyes of a mountain of a buffalo. All my senses were being filtered through the lightheaded breeze blowing across the river valley at 8,500 feet of elevation. All the shapes and colors appeared so sharp and vibrant. The prairie and the mountains seemed alive. I was nearly hallucinating—almost, but not exactly. It was not so much that any particular part of the landscape had dramatically altered—rather, that every part seemed more vivid. I had entered buffalo time.

I do not mean to dwell upon this sensation in order to explain away what happened next. Rather, I offer it as a possible reason why, what did happen next, actually happened that way. I had opened up to the environment. In return, it enveloped me.

There I stood directly in front of the herd, strumming out a drone with my left hand, my right hand weaving patterns of movement through the crisp cool air. Three bull buffalo stepped forward to the edge of the herd's territory. Their bodies were directly between me and the rest of the herd. There they stood, sideways to me, watching, smelling, and listening closely. In turn, the cows and calves ambled along behind them. I stood my ground.

An uncanny thing occurred. I beheld a dirty yellow glow pulse out from the herd, parallel to the three large bulls. It was like a ring, a smokescreen, or a fence expanding outward around the entire herd. It was luminous but it wasn't light; rather like individual bubbles or bundles of glowing energy. It was dots on fire. It stopped just in front of me, like a barrier or signal that defined the group territory: a boundary, a social aura, an energy extension of the herd's group body language. Yet it had no real substance. I felt that it would disappear if I only blinked.

That I actually saw what I believed I saw was in no doubt, because when I put my left foot directly on the ring, the largest bull, who stood less than a hundred feet away, began to paw the ground with his hoof. I pulled my foot away. The bull immediately stopped pawing. To be sure that the connection was what I thought, once again I dropped my foot onto the glowing ring. Once again the big bull pawed the ground.

There I stood, feeling like a guitar warrior staring into the face of the big bull. I moved my right foot in time with the pulse of the drone, waving my left hand palm up to the sky. Now I would step onto the ring, and just as quickly step off. The herd leader looked at me directly, through those enormous sloe eyes of his. I stepped on the ring and he pawed the ground. I leaned to the left and looked at him, then leaned to the right. Intuitively, I felt that he was slowly becoming aware that we were in this dance together, and that it was based on harmony rather than threat.

The sixth time I put my foot on the ring, the large bull did not paw the ground. His eyes locked in on me for a moment, and then he began to move. He kept the same distance but stepped so that he was facing me almost directly head on. His head was tipped ever so slightly, so that he could continue to watch my movements through those big eyes of his. The other two bulls slowly walked away from him and went off to join the rest of the herd. They were all chewing and relaxing, swishing their tails to keep away the flies. At that moment, the social aura, the glowing ring, began to recede inward, back toward the herd. It moved very slowly, almost imperceptibly, like the minute hand on a clock. Just as slowly, I followed it inward. I felt that the alpha bull had come to the realization that, indeed, this two-legged had actually seen the ring, and recognized its purpose. The two of us had invented a dance, a communication. It told the entire herd that I meant no harm. The strange two-legged creature with the strange sounds wanted no more than to dance with them on buffalo time. The buffalo went about their slow business with no further ado.

At that moment, just to test my hypothesis, I stepped right onto the ring again. Just as suddenly, the big bull began to turn sideways and pawed the ground, if a bit halfheartedly. I stepped off the ring, and made no more aggressive overtures. I had learned my lesson. After all, the big 2,000-pound bull stood less than forty feet away. Now, I kept my eyes locked onto the glowing ring, and slowly followed it inward.

Perhaps fifteen minutes later, I was pulled to attention by the sound of a human voice shouting from very far to the rear. I realized that the ring, and the sound, and the elevation had all collaborated to really hypnotize me this time. I turned around. The head cameraman was standing on the knoll waving his arms madly, shouting at me to get the hell out of there. I didn't fully understand why he was so upset, until I noticed a buffalo cow and calf saunter between where I stood and the distant knoll. I was standing smack dab in the middle of a herd of a hundred buffalo.

It was quite impossible to walk back toward the river. Or for that matter, it was also impossible to walk forward or even sideways. There were buffalo everywhere; the larger ones seemed as big as a house. And all their stomachs seemed to be growling, grumbling, and farting at the same time. It was the sound of mortars exploding from five miles away. Yet none of the animals seemed in any great pain. Neither did they take any heed of my presence. There I stood less than ten feet from a large bull who lounged in the grass, chewing and chewing. I still played the drone, still moved my fingers to the gentle rhythm of the music.

The ring had vanished entirely.

The buffalo continued grazing for ten more minutes, certainly no longer. Then they all slowly shuffled off down to the river. Within two minutes more, I was standing all alone.

I walked back to the film crew feeling every step, all my senses still heightened. The cameraman stared at me as I approached the knoll. Each of the men in the crew had a big foolish grin on his face. Then they applauded gingerly. But for the rest of the day, the cameraman kept on calling me a damn fool.

Three weeks later, I sat by the phone, surprised that no one from the television studio had yet called to invite me for the customary preview of the edited segment. Finally I called them, and was told rather brusquely by the show's producer that the footage was absolutely useless. He had gone over that footage for the better part of an afternoon, trying to figure out some way to make it a worthwhile piece for the show. But no matter how he cut the film, it still had the pacing of a snail: entirely too slow for prime time TV. All he could see was this guy with a guitar inching toward a herd of buffalo. An hour later the guy was still inching toward them. It was hopeless.

"What about the glowing ring?" I asked. "Could you see the glowing ring?"

There was a short pause. "There was no glowing ring. What do you mean by a glowing ring? Anyway," he continued, "the film does look good. But we can never use it because it has no payoff."

I did not know that term. "What is a payoff?" I naively asked.

"Well, if only that big buffalo had gone ahead and charged. Not anything serious, mind you, but even five seconds of a little rush would have made a beautiful payoff to the segment. As it is, we wasted a lot of money."

A week later, the secretary for the production company called to inform me that my services were no longer needed. So it goes for those who try to put buffalo time onto prime time.

# 8

# *The Spider Web and the Skyscraper*

*One may gain by losing, and lose by gaining.*

—Lao-tzu

INEVITABLY, ANY STUDY of humans and their relationship to the natural environment must eventually confront the issue of nuclear power and war. The U.S. government has posted evacuation schemes in the newspapers of its major cities—as if we might somehow be able "to get away from it" by driving a few hours out into the countryside. I wonder: Which of God's creatures could survive this human territorial behavior known as nuclear war? Perhaps a few of the blind cave fish who scuttle out their existence several hundred feet underground? Maybe, but the water that flows through the subterranean chamber still drops as rain out of a dangerous radioactive sky. What about the giant squid who zoom through the benthic depths of the Southern Ocean? Possibly, but only if the circulation of deep ocean currents actually takes the inordinate amount of time that oceanographers now presume. Is there anybody home above ground? Who can be sure, although the Antarctic penguin certainly breathes cleaner air than any of its Northern animal cousins. What about all those champion survivors of past technological warfare: the cockroach, the snail, the medfly? More likely, much more likely, if not in descending order. One thing is very sure: Neither I, nor the great majority of any humans I have ever known and loved, will be counted as survivors. We are all irrevocably involved, sucked into the vortex whether we want to acknowledge it or not.

It is a tragic commentary of the times that many of our leaders often discuss this apocalyptic issue the way that our forefathers talked about tornadoes and pestilence. Not only is nuclear war immeasurably different in its scope, but also it is our own creation, a crazy measure of our progress, something we may still be able to control. It is not at all like a tornado. Rather, it is the most technologically complex of machines, in an age where our Specimen School leaders still judge a culture's worth by the complexity of its inventions.

And in a bizarre turn of events, the nuclear issue has emerged as one of the premier rallying points of the developing Participant paradigm. It is a beacon, signaling that we have achieved an impossible level of making war with the earth herself. Likewise, it is a bugle call, mobilizing millions and tens of millions of us to unify in order to take back control of our lives for ourselves and future generations of all species.

More than any other issue of our time, the threat of nuclear war forces us to sit up and look at ourselves as a species, as the human species. One thing the mirror reflects back is that our current self-image is an immediate descendant of the spirit of the industrial revolution. As such, deep within each of us we harbor powerful residues of the myth of human superiority, as well as its separatists' correlates of scientific "objectivity" and reductionist rationalism. We have been exploiting the planet for quite a long time now. In order to prepare the human psyche for its 200-year climb up the technological mountaintop, we needed, first, to set ourselves above and apart from the rest of the natural order. We succeeded gloriously, and yet, it has failed us miserably.

Consider the spider web. It is simple, light on resources, complete to the survival of its creator, geometrically breathtaking, and easily reconstructed in the wake of inclemency. But there are few among us who vouchsafe the spider web as the conscious creation of a creature of equal or greater intelligence than a human being. The point is subtle. We may marvel at it, examine it, but we do not truly respect it. How many times has a construction company halted the construction of a skyscraper because some industrious spider went ahead and built its web on the site? The point may at first seem frivolous. But it is the same issue in kind as the endangered cheetah who gets shipped out to tiny preserves in East Africa in order to accommodate greedy ranchers searching for more rangeland. The image of the spider and the skyscraper is a good indicator of our separation, and thus, our lack of respect for the needs and abilities of other creatures. How far we have drifted away from the philosophy of the Sioux, who believed that all creatures were equally intelligent, thus to be treated with equanimity.

When our Specimen School neurophysiologists determine intelligence in animals, they do so by imposing humanly derived "tests" upon them. A few chimpanzees are captured, shipped in a cage from their home environment, and placed in a laboratory. Then "studies" are levied upon them. After a period of time and interminable repetition, some researcher "draws conclusions." In this particular case it is "proven" that chimps are capable of communicating to human beings via a humanly derived sign language. Chimps are thus deemed more "intelligent" than a host of other contenders including sea otters, octopuses, and our friends the spiders. The many quotation marks employed here are used to accentuate those terms that this author feels do justice neither to their intended meaning nor to the actual purpose of the situation. They are all hyperbole—masks utilized to hide the blind cruelty perpetrated against the chimps in the name of the pursuit of knowledge. In kind, it is the same aggression perpetrated against the spider and the cheetah, identical as the masked reality of the nuclear bomb. The case of the

chimpanzee experiment demonstrates a use of the concept of "intelligence" as we humans have come to use the term for several generations.

Consider your own criteria for intelligence. What is it that you yourself consider to be a sign of intelligence in another? Is it his or her facility to dredge up information to fit the needs of the changing moment? Is it a being's ability to keep a cool head while all about them are losing theirs? Or is it an ability to take command just when some problem arises to impede progress? Does a more intelligent person necessarily get better marks on an intelligence test? Is the atom bomb the creation of a more intelligent creature than the creator of the spider web? If so, then is it better to be more or less intelligent? Is it all just so many apples and oranges?

Certainly the concept of intelligence involves such ideas as learning from experience, reason, and the construction of abstract concepts. Yet here and now I defy anyone to put forth a case of any such act that proves that humans are more intelligent than any other animal. We humans can write, program computers, poach eggs. Is this proof? But then the plover and the monarch butterfly can navigate several thousand miles by reading the earth's magnetic field. Eels may know and use electricity as well as any engineer. All these examples may be construed as species-specific attributes. Then there is the issue of *simplicity.* Is an animal less or more intelligent because it lives without clothes, central heating, and, well, atom bombs? On the other hand, one might easily argue that the relativity theories of Einstein are an exclusively human province, a profound proof of human intelligence. But I wonder, how can we know whether or not those same monarch butterflies employ some relativistic compensation in their search for the same leaf they left last year?

No doubt, some of you believe that I am treading on very thin ice. Others of you are probably convinced that I've already fallen in, and that the prolonged immersion has left me permanently addled. I do not add this apology as a disclaimer; rather, it is a conscious attempt to pull you deeper into this train of thought by personalizing it. This author is doing nothing more than attempting to understand how this wonderfully complex human race could have evolved to this insane impasse of nuclear proliferation. Regard the chameleon. Its intelligence has worked out a decision-making process of changing skin color from green to yellow to brown to black. Its unerring quick-wittedness to choose the correct suit for every occasion has helped the chameleon to survive intact for a hundred million years or more. I cannot help but wonder if, in the big picture of life on earth, any individual chameleon possesses more intelligence than, say, some politician who votes for increased arms production in hopes of winning more personal clout within the Washington or Moscow military establishment. And for that matter, if the only rhetoric that begins to make any sense out of this absurd situation bases itself upon an equation where survivability equals intelligence, then let us all pay our respects to the mitochondria—at the head of the class for that unique biological distinction of having survived intact for more than a billion years.

Granted, the survivalist's approach to intelligence does encounter some serious problems. Intelligence is different than instinct—it is different than simply making the right choices. Intelligence implies a certain loftiness, as if it was never meant to do anything more than define a hierarchy, with human beings located, ipso facto, at the peak. Somehow the concept of intelligence simply does not apply to mitochondria. Perhaps the single-celled organisms do indeed possess some still unknown qualities of consciousness. But not intelligence. In fact, our education tells us that intelligence barely applies to chameleons either. A chameleon may *choose* a specific skin color to elude a hungry predator, but it never *CHOOSES* its skin color. Despite what we may feel about the choice of the two examples given here, that chameleon is no Pentagon general pondering various nuclear evacuation schemes in order to *CHOOSE* the best plan.

Back in 1863, Thomas Huxley developed a theory that stated that animal species demonstrate a continuous gradation of mental capacities similar to their gradations of physical structure. In typical nineteenth-century style, Huxley equated complexity with intelligence. Also in keeping with his time, he concluded that human beings were the most complex physiologically, and thus the most intelligent of animals.

Much more recently, H. J. Jerison proposed that encephalization—the ratio of brain volume to a body's surface area—reflects a more logical criterion for measuring intelligence in animals. This view is entirely anatomical, and never really addresses itself to either intellect or consciousness. However, Jerison does seem to imply that they all go hand in hand. The human being sits very near the top of Jerison's hierarchy.

But most significantly, the human being is not the number one encephalic. That honor goes to certain species of dolphins, with *Tursiops truncatus,* the bottlenose dolphin, at the top. In a way, it seems especially ironic that most of us have come to know the bottlenose dolphins from their role as clowns in oceanariums. Shakespeare may have been correct after all: It is the Fool and not King Lear who best perceives the truth.

The dolphin brain evolved to its present capacity at least 15 million years ago. That also means, of course, that 15 million years ago the dolphins had evolved to a point of intelligence that human beings have not yet arrived at today. They, too, must have searched into the many realms of consciousness. They, too, must communicate some understanding of the universe to their children. It may include many things about the nature of life that we humans cannot even begin to imagine. John Lilly, an important transitional figure between the Specimen and the Participants schools of animal researchers, has studied the bottlenose dolphin in captivity for many years. He states categorically that the dolphin is more intelligent than humans. Furthermore, the expression of that intelligence is very different than what we humans *think it should be*. Lilly has concluded that the only way to decipher the intellect of the dolphin is to bridge the interspecies communication gap through a mutually derived language. Once the code is cracked, the flow of information between humans and dolphins will initiate a paradigm shift in human consciousness on a level with Copernicus's heliocentricity, which demonstrated that the earth is not at the center of the universe. Likewise, the dolphins will show us that we are not the only intelligent life on earth.

Yet, it is at the juncture of studying the mysterious mindset of the dolphins that we encounter a profound impasse. Whenever we examine the accomplishments of the dolphins using the yardstick of our cultural and technological bias, we always come up empty-handed. This particular metaphor seems especially potent, for the dolphins themselves are "empty-handed." They have no facility or need for tool use, which is an attribute of hand-eye coordination. In a way, the dolphins are living proof that tool use itself is no criterion for measuring intelligence.

But if Lilly's paradigm shift is imminent, it still does not exert much influence in the circles where power is wielded—where tool use, and intelligence, mean fundamentally the same thing. And it is from this territory that the bombs continue to be manufactured at a fever pitch. After all, the bomb is one of our most complex tools. In a bizarre way, one may apperceive it being held up by the superpowers as a symbol of the upwardly striving superiority of this or that national intellect. It is as if we were to pat ourselves on the back for painting ourselves into a corner.

The image is impossible to ignore. And most importantly, it is at this juncture that many of us have begun to make the leap into a new frame of consciousness. If this is intelligence, then a fast-growing number of us don't want it anymore. Yet it is not that we are Luddites—naively anti-technological.* Rather, we promote an appropriate, pro-life technology. Rather, we turn to the famous statement made by Ishi, an American Indian who was captured and later educated at Berkeley around the turn of this century. When asked about the technological achievements of modern civilization, Ishi replied: "You people are smart, but you are not wise." Or consider this, a statement by James Thurber:

> I observed a school of dolphins . . . and something told me that here was a creature all gaiety, charm, and intelligence, that one day might come out of the boundless deep and show us how a world can be run by creatures dedicated, not to the destruction of their species, but to its preservation.

On an island in the Gulf of Carpenteria in northern Australia lives a tribe of aborigines known as the dolphin people. This tribe has been purported to be in

*The Luddites were a group of workers in nineteenth-century England who smashed new labor-saving textile machinery in protest of a loss of jobs. Often used to refer to someone who is anti-technology.

direct communication with the wild dolphins who reside just off the coast, and for many thousands of years as well. Their shamans remain heir to a complex series of whistles that signal the dolphins to draw close. At some point, the whistles dwindle and finally stop altogether. The shamans explain that they begin to speak with the dolphins mind to mind. We place the aborigines among the most primitive of human cultures for any number of reasons. Yet they are easily able to survive in the desert where the more civilized among us would surely perish. Now, their culture appears in a process of rapid deterioration owing to the inevitable civilizing influence. The elders describe the modernized aborigines as a people who have lost the essential attributes of a healthy, attuned mind. How ironic that we label as primitive what the shamans denote as healthy.

One might extrapolate that according to Ishi, to Thurber, to many of the Participants School of zoologists, to the aborigines, and most probably to the dolphins as well, our hierarchical definition of intelligence must be diagnosed as unhealthy to the organism.

While the keepers of the technocratic society spent a few recent centuries building and then destroying their handiwork several times over—and then finally stockpiling fifty thousand nuclear bombs in the mad assumption that it would protect us from our enemies, enhance our chances of survival, as it were—all the animals from mitochondria to dolphin to earth-focused human beings everywhere lived and died in close harmony with the currents of the earth. Yet let it be plainly understood—it is not that humanity has turned out to be the bad apple in the bunch, but rather, that our exploiting attitude, our mindset of superiority, was a disease. Now we need some strong medicine. In a macabre sort of way, the bomb has turned out to be one of our doctors, sternly pointing out the symptoms at a time when we need to know the facts. Curing ourselves will take a concerted effort.

One major part of that cure involves realignment with those currents of the earth. As such, we must rid ourselves of such unhealthy concepts as the one known as "intelligence." It is blatantly counterproductive, in fact specieist, and is inapplicable to the concept as it actually exists in nature. But whatever the word may be, I define it as any creature's ability to utilize its own unique functions of mind to emulate the success story of the mitochondria. Regard the dolphin. He or she musters all that colossal intellect in a one-pointed attempt to devise a "thinking man's" version of the wisdom of the one-celled creatures.

When we allow ourselves to dwell within the mechanism of this natural ethic—this ecology—then everything under the sun begins to appear as a unity. There is no more intelligence, and no less, for the turtle, the raccoon, the hermit crab. The separatist hierarchy known as intelligence has now metamorphosed into the earth unity known as *natural wisdom*.

There is a punchline to all this, and it gives this story an ending with an ironic twist. You see, not only have we humans *evolved* from the one-celled creatures, our elders as it were, but also, we have *incorporated* them directly into our own bodies. At this very moment there are billions upon billions of quite independent mitochondria going about their business within each and every one of us. It is they who control the intake and distribution of oxygen to each individual body cell.

This clear presence of an independent organism within "our own" cells easily confuses the existential issue of self-identity. Am I an inflated bag controlled and manipulated by swarms of mitochondria; or am I me? Microbiologist Lewis Thomas settles the issue flatly by declaring, "They run the place." By some kind of intellectual osmosis, we "ourselves" are our own elders. We already possess the natural wisdom.

Human culture has always treasured its sages. They sit calmly in doorways, pointing out the spider webs and skyscrapers to our children. They always seem best able to express themselves through laughter. It is our wise men and wise women who have always promoted a life of spirit, showing us how best to align the unities of life with our day-to-day struggle. It was they who discovered those great metaphors for survival, and so named them "love" and "simplicity" in any of a thousand different languages. Now they are mobilizing the rest of us around the issue of peace. They have already convinced many of us that in unity we have the power to survive.

# 9

# *What the Orca Says to the Seagull*

## I. Sounds in the Water

WE WERE CAMPED in Buddy's Cove for nearly a full week before I got any practical opportunity to play music one-on-one with the orcas. Oh, there were plenty of whales around. That was not the problem. In fact, at least one person in our party of eight sighted either an individual orca, or even a pod of twenty or more, three or four times a day. However, we had not traveled so far up into the British Columbia inland waterway wilderness merely to *see* orcas.

I was feeling grumpy, a bit deprived, frustrated to boot. I sat around the rocky, slippery shoreline, eyes peeled to the swiftly moving straits, listening for a sign that today would present the proper circumstances for an interspecies musical improvisation. Too many noisy boats. *Brrrrrrrrrrrrrrrr, clug—clug—a-clug*—all day long. Instead I'd pick up a loose piece of the ubiquitous driftwood that littered the shore, and begin carving convoluted heads, hands, whales, wolves. Whatever shape the driftwood took by itself, that is where I'd go with it. If it was flat on one side, all the better. Then I would incise oval eyes. Big full lips wrapped around a toothy grin. It was no secret to me what prompted the Northwest Coast Indians to become such imaginative wood carvers—a mix of boredom, frustration, and an endless supply of shapes waiting to be sprung from soft driftwood. It was a creative adult's reply to the child who spends all day throwing stones in the water. My first wooden creation was a kind of Muppet-saber-toothed tiger with huge grinning incisors. He was completed in a morning, and promptly mounted on a pole that was stuck between two boulders next to the food fire. Next came the wolf-dolphin. Then the King Tut-chinned orca. Within four days' time there got to be so many wooden faces peering toward the cove that finally our cook pleaded with me to find some other occupation. The campsite seemed too much like the upper tiers of the Notre Dame cathedral, with all its gargoyles peering out over the people of Paris. When I started walking around camp like the hunchback of Notre Dame, she threw a rock at me. Nothing large, mind you, but it gave me a great idea. For the next two days, I spent hours throwing the flat stones into the waters of the cove. Just to watch them plop.

Once or twice a day I'd join my companions out in the straits that abutted Buddy's Cove. We'd scrape our fiberglass kayaks over the stones and up to the water's edge, climb into the cockpits, and gracefully slide away over the glassy waters of the cove. There were giant cedars to stare at, magnificent rust-barked giants as old and large around as the more celebrated redwoods. There were mountains, distant glaciers, waterfalls.

And of course there was the wondrous joy of paddling the kayak. It never seemed to matter where we went. It was the fluid, liquid motion of the voyaging, the arms pulling and pushing in synchrony, slicing the bow through the still, deep waters. Kayaking is more like hiking than sailing. You walk with your arms over the water, moving across the landscape by your own power. Within minutes we'd all be sitting silently a mile offshore, face to face with one of the most luxurious wilderness views in the entire world.

To my one-track mind, the most exciting aspect of the landscape was the one I could not see. Evidently, the waters adjacent to Buddy's Cove drop off very quickly to more than a thousand feet, and just four hundred yards offshore. It is easy to picture a dark, obscure, craggy bottom, populated by creatures in grottoes, far more fantastic than even a Muppet-saber-toothed tiger. If I ever get the opportunity to "crack the code" of orca language, my first question to them will be about the denizens of this offshore trench.

We glide through the waters for more than an hour. The sun hangs low over the mountain; we head for shore. There! Someone spots the enormous wedge fin of a bull orca. They are a good mile farther out in the strait, six of them heading south to one of their favorite feeding grounds. Yesterday three of them, including a ten-foot-long toddler, swam right up beside my kayak, as if chiding me to race them up the coast. I didn't take up the challenge. I was too excited just to watch them swim upside down, white belly up, and then surface two hundred yards in front. One of the bulls appeared ten feet longer than my seventeen-foot kayak. Then they all blew together like so many muffled cannons: *puuuu-puuuppuuu.* Three up and then three down. Suddenly, a forty-pound salmon broke the surface just in front of them. When the orcas are about, the salmon seem to jump higher out of the water than seems possible for a creature without wings.

Day after day I lolled around the camp, watching and listening for an opportunity to try my luck at interspecies music making. And day after day something would interfere. The noise. The major problem was the noise. The area around Buddy's Cove is just teeming with boats. There are commercial fishing boats, private outboards, ocean liners, whale-watching charter boats, scientific inflatable boats, and barges. In terms of sheer noise, the barges are by far the worst offenders. They are eighty feet long but sound ten times that: square, powerful tugboats pulling two and sometimes three massive log islands. First you see the boat like a steel ox plowing through the strait. Then there is a thick steel cable attached to a thick floating fence that contains three or four acres of logs. Then another long cable and another island. The barges chug through the straits slowly, *clug—clug—a-clug*, ever so slowly and purposefully, taking a full hour from the time we first hear them to the time they are gone. Then, finally, silence. The underwater environment reverts to its crackle static of shrimp, the low whoosh of current streaming through kelp, the strident soprano saxophone cry of the orca.

To monitor the underwater environment, we use a type of underwater microphone known as a hydrophone. As the days pass one into the next, it has become positively disconcerting to realize that the cumulative noise of all these waterborne internal combustion engines has turned the otherwise serene wilderness into a kind of hidden urban freeway. It is hidden because its effect is almost entirely underwater. Most of the visitors who zoom through the area have absolutely no conception of their own uproar. We humans exist above water, perceiving the environment mostly through our eyes. We come to a place like Buddy's Cove and feel great tranquility at the sight of panoramic wilderness sweeping right down to the water's edge. To any human observer, the boats appear like inconspicuous little toys, lost in the immensity of the setting. Even the log barges pass unnoticed by a group of campers immersed in the busy task of eating a meal before a campfire.

But underwater, where the whales perceive their environment through an exquisitely sensitive sound-processing system, humankind stands guilty of polluting the wilderness in the most harmful manner possible. It is partially a matter of physics. Sound travels through water four and a half times faster than it travels through air. This property necessarily foreshortens the acoustic horizon. You can hear the sound of a boat underwater long before you can either see or hear it above water. Whale acousticians have postulated that some of the great whale species may be able to signal their own kind over distances measured in the hundreds of miles through a combination of very low frequency and very loud volume. That may have been true at one time, perhaps a hundred years ago, before the proliferation of powerboats. Sitting on the shore of Buddy's Cove, listening to the toothache throb of an invisible barge, I try to imagine how the drone of powerboats has changed the communication abilities of this resident orca population. I wonder how an orca would define the word "progress."

Unfortunately, the noise is not my only problem. Music demands attention—a mutual acknowledgment in time and in place. A musician cannot play in a band unless the various instrumentalists become mindful (heartful?) of one another's sound. All together, the band makes magic by channeling harmony. Separately, they make cacophony. Here in Buddy's Cove, after a very frustrating week of carved spirit watchdogs, of whale vocalizations drowning in a drone of motor noise, I have felt no real chance to catch the orca's attention. "Hey you orcas, when you got a free minute, why not drop by my place."

The trouble is, they just don't have the time. Every orca pod seems to be closely followed by some daring young man in an inflatable boat. The boats, generically called Zodiacs, never seem to be more than a hundred yards away from any pod between the hours of 8 A.M. until 5 P.M. When we ask the people, in our naiveté, what they are doing acting like so many pets on a leash, they all answer with the ominously stock reply: "We're studying the whales." Some of these students are aligned with universities; others are filmmakers. Some are well along the path of earning postgraduate degrees in behavioral puzzle solving. Some are overly purposeful vacationers who have recently learned that the orca is not the bloodthirsty predator it was once purported to be. Rather, it is some kind of animal intellectual. People who last year trekked in Nepal are this year studying orcas.

My, oh my, it is easy to get cynical sitting around this damn rainy campsite all week long.

The same scene is repeated all day long. A Zodiac zooms into view with a cinematographer strapped securely to the bow. His camera peers straight down into the depths, attempting to film valuable close-ups of the orcas who cruise just below the surface. The camera catches a green spot gliding underwater, which very gradually transforms into white as it breaks the surface to be identified as an orca's cheek patch. Unfortunately for the whales, this ten-mile-an-hour exercise in cinematographic perseverance may continue for several hours at a time. They say that any filmmaker worth his salt will take at least ten shots of the same subject in order to achieve a single usable print. I know all this firsthand. Two years previously, I had been a member of such a film crew. The experience was unsettling. I prided myself in possessing a sensitivity to the needs of the whales. A filmmaker is a harasser. I was a harasser.

This morning, the film crew of the day pays a courtesy visit to our campsite, no doubt curious about my intentions. Genuinely interested, I ask the director if any of them has yet hit an orca with the razor-sharp propeller of their ten-horsepower outboard. He peers at me through intense eyebrows, and then answers in a hurt voice that the orcas can assuredly judge their distance from the whirling blades. He asks me if I've ever seen dolphins riding the bow of a boat. Yes, I have seen that, and indeed, sometimes the boats go much faster than his little rubber Zodiac. However, in that case, the dolphins are not always willing participants here. They are not permitted to leave when they're too tired to make snap decisions.

The filmmaker documents whale activities because he believes that the image is monumentally beautiful. He works hard to capture that animistic wild beauty for other people to share, if only through the secondhand vision of the movie screens. He and I both know that the long propeller scars seen on so many of the local whales are certainly the handiwork of strangely motivated humans who had intentionally swerved their boats at the whales as they surfaced to breathe. Perhaps that too is an expression of humanity's relationship to the contortions of anima.

The two of us sit facing each other on long ocean-bleached cedar logs, no doubt contemplating the terrible image of propellers, both staring intently at our hands. People in wilderness talk to each other in a different cadence than people in cities. We share the wild silence as equally as we share each other's words. He asks me if I eat meat. "Yes, I do," I answer, but only once in a while, and then mostly as a condiment, "like the Chinese."

The image of a wounded orca looms back into consciousness. Both of us are amazed that the orcas have never been known to retaliate against such malicious violence. Instead, they dive very deep, only to surface very far away from their would-be assassins. Oh, there is a legend or two of an orca waiting patiently to later enact a crunching revenge in its huge toothy mouth. But no one we know has ever heard or seen such revenge firsthand. It is at this point in our discussion that the inevitable truism is breached: "How strange that *we* call *them* killer whales."

I mention nothing to the filmmaker about the hidden rudeness of noise pollution. Instead, I encourage him to refer to the large film library of another filmmaker who also spent weeks chasing the whales around the strait. Rather than repeating the same old process over and over again, instead, why not share the accumulated treasure of stock footage? After all, 90 percent of the action is what I label the "whale-cruising-blowing-wonder-and-awe shot." This particular filmmaker shakes his head. He is hoping to catch something unique and unexpected. I ask him if he would be willing to share that footage with future filmmakers, even the unexpected stuff. He lifts his head and looks at me through those intense brows. The look itself informs me that he does not believe that his one-week filming jaunt with the whales is detrimental. Neither can he fathom the fact that there will probably be three or four more film crews here within the next three months. Then he answers my question: "Of course I would." Unfortunately his tone is not very convincing.

The film crew lounges around our warm campfire for the better part of the afternoon. We start out by drinking cocoa, but by the second hour we've switched to straight tequila. By midafternoon we are joined by another man who also arrives by Zodiac. "No," he tells me as he pulls his boat high up onto the beach, "this is not a Zodiac, this is, thank you very much, an Avon." I cannot tell the difference. To my kayak-oriented mind, Zodiac is a generic term, like Band-Aid and Jell-O.

The newcomer introduces himself with a flourish. He is J.S., the resident orca scientist. This year marks J.S.'s fifth continuous year of study with this specific population of orcas. He knows their movements, their habits, and their group dynamics better than any other human. J.S. should have been a comedian. He speaks in that self-parodying manner that puts quotation marks around key words. Sometimes he actually raises his hands and flutters his fingers as if writing the quotes as he speaks. He announces to us that he comes from the good-ol'-boys school of biology, where an animal is truly a "dumb" animal. "I 'know' more about these orcas than they 'know' about themselves." The twinkle in J.S.'s eye makes it only too clear that he is spouting an official party line—the science party. Maybe so, but he can't fool us. It is obvious that the man has witnessed some pretty bizarre behavior from these "mere animals." We've all felt it. These orcas, more than any other animal, make one a believer in that old American Indian view of nature where everything is conscious and articulate. But whatever he's seen, he's not telling. The bottom line for J.S. is that behavioral information about the whales is a kind of product, to be bartered in obscure journals for academic credentials. One side of him cares very much more about this product than about the witnessed reality of the orca. He loves to spend his summers making this kind of product. He loves the whales the way a professional skier loves a tried-and-true pair of skis. They are beautiful, and provide him with an active outdoors lifestyle with pay. Of course they also provide him with an ongoing "mystical" experience, again like skis, but J.S. definitely prefers to keep that kind of knowledge to himself. In a way, he reminds me of a dreamer who wishes to sublimate his dreams.

J.S. knows every local whale by name. He has a thousand anecdotes about how A-5 (number five whale in A pod) did such and such with A-7 this morning. He sits down next to me on the cedar log and proceeds to spend twenty minutes describing how A and B pods got together last week. It all sounds like an interoffice softball game to me, with J.S. as the sportscaster. Maybe that is my problem with his scientific method. It is too eye-brain oriented, like a sportscaster who spices up a blow-by-blow account by bringing in statistics from past events. I want to know what J.S. *felt,* witnessing this event which is evidently quite rare and formalized. Instead, he rambles on and on. It is a great story, and J.S. is a great announcer. The pursuit of knowledge as an ongoing series of witnessed events.

He is an ethologist. That is, J.S. studies animal behavior. This year he is very hard at work cross-referencing the number of breaths with various behavioral patterns. He announces to us all that today we might be overjoyed to know that each animal in his group of ten breathed an average of 162.7 times per hour.

We all laugh at the way he describes his work. I want to know how such information can possibly be of use to anyone. He looks at me with his characteristic twinkle, downs a shot of tequila, winces once, pulls at his very real mustache, and finally answers in a heavy Mexican accent: "I am helping the whales, señor, by expanding our knowledge of their behavior." Silence. J.S. puts on a good show, but I just don't get it. Maybe I'm just too grumpy. If anything, it reminds me of the government spying on its citizens as a way of keeping the country free. J.S. knows all of this, of course. "Just remember," he adds in his normal voice, "ten years ago, whale scientists were killing whales to count the average length of their penises." Then he twirls his mustache again. "And remember, señor, it is the only way I can get the master's deeee-greeee."

The filmmaker has had enough of J.S's mannered patter. He looks at me and asks about my aspirations of creating an interspecies music with orcas. I look him in the eye, and then shift my glance over to J.S. What a strange trio we make. Three different fields of human endeavor sitting around the campfire slurping tequila. All together we tell the story of the human/orca connection. Each of us somewhat guilty of harassing whales just by being present at this moment in Buddy's Cove. All of us loving the orca in our own way, each of us critically wary of the other's presence. Each of us only too aware that there are too many people this summer in Buddy's Cove. And what of next year? The filmmaker views the orca as his subject. The ethologist looks at the same being and sees a specimen. What about me, the musician? I hear a player. I suppose I see a neighbor as well. If I tend to get self-righteous with the filmmaker and the ethologist, it is because I truly believe that it is better to treat the whale as a neighbor than as either subject or specimen. But there is no need to orate. No one here disagrees with me. As men, we share a deep spiritual unity with orca. As professionals, we get wary, self-righteous, and self-parodying.

I pull out an electric guitar and pick out an arpeggiated D-major chord. The thin sound of an unelectrified solid body fills the quiet campsite. The D chord competes only with the crackle of burning driftwood. Then I switch to a G chord and play the same riff a fourth step higher. Then it's back to D for a while, up to A, and finally back to D again.

Once, two years previously, I had played that same progression through an underwater sound system and into the water for three nights in a row. All three nights the music attracted orcas. When I started playing, the whales could not be heard. After ten minutes of playing, the orcas began vocalizing, but from several miles away. Fifteen minutes later they arrived at the cove, and spent the next three hours interacting with the guitar music. On the first night it seemed as if the whales vocalized constantly, not at all coordinated with the harmonic and rhythmical structure of the chord progression. On the second night, the whales arrived as I was getting my equipment ready to begin. One individual whale stepped out to take a kind of lead voice with the guitar playing. The rest of the pod chose to stay in the background, jibber-jabbering among themselves in a quieter tone, which seemed unrelated to the unfolding ensemble playing at center stage. At the same time that the whales split into singer and Greek chorus, a group of humans appeared at the seaside sound studio, no doubt also attracted by the guitar/orca music that was being monitored over some hi-fi speakers. They, too, began to comment among themselves at key places in the interaction. Sometimes the human observers would comment at the same moment that the observing orcas seemed also to comment. Once, the correlation was so clear that I had to stop playing a moment, just to get my bearings. Other times, there would be no relationship at all. However, on this second night of music making, the orca singer was still not able to match my tune very well. I had intentionally chosen a very tricky reggae rhythm, so that if the match was made, there would be absolutely no doubt about it.

The third night evolved into pure magic. The session began at 10:30 P.M., the same time as the two previous nights. And as the night before, the whales were there promptly at 10:30. In planning the session, I had first considered scheduling it more in accordance with the tides—an hour later each night. But somehow, I never followed through on that notion because both the whales and I were ready to play at 10:30. I began the session by mimicking the standard stereotypical vocalization of the pod: a three-note frequency-modulated phrase that begins and ends on the D note. But this pattern is never frozen. Rather, it varies in form by the addition or deletion of the speed of the glissando, by the fluidity of the legato. In other words, the whales' language varies just exactly the same way that a jazz musician varies a standardized melody. And the whales seemed very aware of my attempts to vary their song by ending each of my phrases with a solid obbligato amen of D to C to E to D.

Unfortunately, the highest note available to my electric guitar is a mere C-sharp, an impenetrable half-step universe below the orca's tonic note. Thus, in order to reach their register, I needed to bend the high string—something ordinarily not that difficult—but, in fact, rather clumsy to achieve hunched up in the dark fog while fingering up at the very top of the guitar neck. The first time I attempted the bend, the result sounded like a very respectable approximation of the orcas' phrasing. Actually, to be totally honest about it, if the whales had been saying, "Try again, you ignoramus," then I replied, "*Twa wawa woo inwawawa.*" But I chose to remain undaunted by such matters as content, and so repeated the phrase a second time. Suddenly, the high E string snapped. While I sat there in the thick night air fumbling through my guitar case for a fresh string, the orcas stepped up the intensity of their vocalizations. Calling, calling for me to rejoin the music. Every so often one of them would punctuate a long sinuous phrase with the obbligato.

Utilizing the language of my musical training, it feels very comfortable to name such an encounter a jam session. Furthermore, the term "jam session" evokes such other constructs as improvisation, jazz, even that large human cultural art form that we call music. *Music!* Behaviorally the orcas may use their whistles as a kind of signature to keep in touch with one another over a mile or more of murky waterway. With me, they were using those exact same whistles to invent melody, rhythm, and harmony.

But perhaps I stand guilty of bald-faced anthropomorphizing. In other words, in order for these signature whistles to be called music, must not the orca hold a concept that is at least analogous to what we humans know as music? I disagree. What we invented was neither human nor orca. Rather, it was *interspecies* music. A co-created original.

Thomas Sebeok, a student of animal signals, has written that animals certainly possess communication, but that not one other species besides human beings possess language. Thus, it is no easy task to step from the intuitive, and thus scientifically dubious, declaration of musical collaboration directly into that ponderous, concept-bandying dance that we English-speaking humans call "language." Certainly music is communication. Any musician would argue that music is also much more than that, that it is also a complex and concise language. A poet or a psychologist might call it the language of emotions; the priest might define it as the language of communion. It is arguably as profound as and more universal than any individual human tongue.

But even if music is the most profound human language, still, it is not *the* language. *The* language is English, or Japanese, or Swahili, or any of a thousand other lingual structures that coalesce humanity, yet prevent humans from comprehending one another. The definition of *the* language involves half a dozen academic disciplines. Music is more easygoing. It is defined by the listener. More so, it is not so tied down by the knottiness of several competing schools of linguistics, behavioralism, education, and psychology—all ready to fight it out for ten rounds as soon as someone states he or she is communicating through language with an a-n-i-m-a-l. I wonder, how do you say "animal" in Aborigine? Or to confuse the issue still further, how do you say "a D-natural note" in Orca?

At that particular moment in time, I felt no particular desire to step on anyone else's hard-fought definitions. Instead, I tightened up on the E string, and stubbornly plucked out the orca's obbligato, but this time in C-sharp instead of D. The center-stage orca immediately answered by repeating the phrase in C-sharp. Otherwise, it was the same exact melody. From that point on, the dialogue between us centered on the common C-sharp chromatic scale. And the conversation continued for more than another hour in very similar fashion. By that time, the interspecies connection had attracted eight more people, all crammed into the shore-side camp to listen to this exchange of information and emotion between whale and human. At first, the human audience deemed it necessary to comment each and every time that any one of them discerned a recognizable pattern within the exchange. This seemed surprising to me since neither the orca nor I was focused upon any set melody or rhythm. The connection was one of feelings and communion, making itself known by an ancient trait of music that transcends intellectual patterns.

Instead, what the orca and the guitar player had settled upon was the conversational form of dialogue. That is, each of us waited until the other had finished vocalizing before the other one started. In order for such a form to work properly, both of us had to become acutely conscious of each other's beginnings and endings. Once in a while, one of us would inadvertently step out before the other one had completed his piece, but in general, the form of the dialogue was clearly working. And as such, the resultant musical exchange never digressed to a mere call and response. Thus it was very different in both form and spirit from the response of a mynah bird or a parrot. There was always a feeling of care and of sensitivity, of conscious musical evolution within the time frame of a single evening's music. I might play three notes and the orca might repeat the same progression back to me, but with two or three new notes added on to the end. Once, I made an error in my repetition of one of the orca's phrases. The whale repeated the phrase back again—but this time at half the speed!

After an hour of this intense concentration, I truly felt as if I was going to explode. There was nothing else to do, no place else to go with the dialogue but directly into the sharply etched reggae rhythm of the previous two nights. I played it, inexplicably, in the key of A. The orca immediately responded with a short arpeggio of the A chord. When I hit the D triad on the fifth downbeat, the orca vocalized a G note, also right on the fifth downbeat. It was the suspended note of the D triad. Then back to A and the orca responded in A, again on the downbeat. The agile precision of rhythm, pitch, and harmony continued through the entire twelve-bar verse.

The second time I played the song through, I made several glaring errors in both melody and rhythm. Likewise, the orca's performance was entirely lackluster. I cannot speak for the whale, but I know that for myself, I felt some mad desire to make this second verse still more complex and tricky. Yet all the signs cried out for simplicity and further relaxation of the humanly imposed structure. Naturally, my fingers could not figure out whether to follow the needs of my head or my heart. What had commenced as a heart-to-heart musical invention had suddenly imposed itself upon my brain in glowing neon as: penultimate interspecies communication. What had been free jazz and then reggae had instead become intellectual revolution and possibly even "historical." One moment I was a rhythm guitarist providing a strong backbeat for a hot soloist; the next moment I was documenting an awesomely successful example of human/animal communication. Under such weight, the music collapsed. On some very basic human level the achievement rubbed up too hard against all my preconceptions of what it means to be human. If I was to grow into this interspecies dance, then I had to concentrate more fully on being, rather than on doing. I had to become more the child.

As I say, I cannot speak for the orca.

## II. Tales of the Orca

Extraordinary tales about orca, the killer whale, abound among the human inhabitants of this rugged British Columbia coastline. Such stories might sound like the foundation of a modern mythology were it not for the fact that many of them are documented. They involve motorboats, movie cameras, patient observation, and living, breathing whales, who nevertheless possess powers that are sometimes difficult to acknowledge. Among the so-called "whale people" of this area around Buddy's Cove are many who have witnessed firsthand the full force of these powers. This explains the storyteller's faraway misty look, a look directed into a magical dream world where giants dwell and all creatures share a common and very ancient language.

For example, there is the story dutifully reported in the log of the Orca Research Center located just up the coast from the Cove. For years, the resident scientists had been espousing the concept of an orca "embassy," a meeting ground where anyone with an idea for contacting the orcas could try it out in a wilderness setting. Late one night, caretakers Jim and Muffin O'Donnell were sleeping in the hand-hewn cabin built right next to the water. Jim was suddenly awakened by a loud trumpet-like sound issuing from the direction of the strait. It was the orcas vocalizing directly into the night air. Very rarely do they sing above water. For what reason would they want to do it then? Jim climbed down the slippery rocks to the water's edge. There, reflected in the beam of his flashlight, he witnessed a pod of orcas staring up at him, and very close to the shore. In Jim's own words:

> I play the flute, can you hear me? I take out the flashlight and make a fast circle on the trees. Will you come near me? I sit on a log and close my eyes. "If you hear me, please come near." My energy amps up, the orcas are very close by; maybe there are four, my eyes are still closed. I hear them breathing, laying on the surface close by the rocks. One of them has a strange sound after each breath—like a long low growl—like something is wrong. I shine the light on me, stand up and walk to the edge of the water. I hold up my hand and shine the light on it and on my face. The orcas stay very still except for the loud breathing. Then I feel a conversation in minds or mind. I am not sure what is happening. I seem to remember a short discussion and apology. Whiz! What is happening here? I felt good. I felt bad. Something is wrong with that one. I left it with as much love as I could get together.

The next morning a group of scientists studying eagles drops by the lab on their way out into the strait. They warn Jim about a boatful of boozy hunters, whom they spied yesterday, shooting at the orcas. When the scientists motored close to protest, the hunters had threatened them as well. Suddenly, Jim's encounter with the orca becomes startlingly clear. That whale with the growling breath had been wounded with a bullet hole in its lung. The whales had visited him last night seeking help and sympathy. After all, Jim O'Donnell and the Orca Center have always received the whales with respect.

Jim calls up the Canadian Mounted Police. The police, however, are slow to respond to the complaint. Taking potshots at federally protected orcas is a very low priority item on their busy schedule. But they also raised their guns and threatened the eagle scientists, Jim tells them. *That* must be worth investigating. Yes it is. A police boat is dispatched; the trigger-happy hunters are located, and told to go home to Vancouver before they are cited for a felony. But Jim is still upset. He writes:

Now what? Incident forgotten? And where is the orca with the long low growl? I am sorry, please forgive our ignorance.

Then there is the story about the dog, almost implausibly named Phoenix. Michael Wills is a local salmon fisherman who takes Phoenix out with him every day. One day, while tending his lines, he realized that Phoenix was no longer in the boat, although just a moment previously he had heard the dog barking loudly. The man brought in his line, and immediately returned to his waterside home, mooring his boat in its customary spot just offshore. A few minutes later, seated in his kitchen, Michael sighted a pod of orcas heading directly for the small cove in front of his house. The orcas stopped just in front of the waterline rocks, and there, deposited a very wet but alive Phoenix. Then they quickly about-faced and headed out into the strait again.

Or still another myth—the legend of the overanxious Japanese photographer who journeyed all the way from Tokyo to Buddy's Cove to photograph the whales. Inexplicably, the orcas, usually so abundant, were nowhere to be seen. The man motored out in his Zodiac every morning before dawn, and spent the day cruising the strait back and forth searching for orcas. Nothing. Then on his last morning, the photographer was all packed, sitting on a stump, waiting for his floatplane to arrive. Suddenly a single orca appeared in the cove and swam right up to where he sat dejected on the beach. The whale actually swam halfway out of the water, so that its white belly rested on the beach pebbles. The photographer jumped up, and rushed around shooting roll after roll of film. Then the whale abruptly turned about and swam away. The floatplane arrived. Later, the photographer would comment that he had been visualizing that exact situation when the orca arrived.

Jim's story was related to me firsthand. The Phoenix account was heard secondhand. The photographer's story was firsthand. Now you, the reader, have heard all three myths, one generation further removed. How might they change in the retelling? All three accounts share one common thread. They make orca seem an imposing candidate for the Nobel Peace Prize. So for the moment, to achieve the proper balance, let us forget this being, orca, and instead focus our attention on this other creature we call the killer whale.

A group of human observers notices a thirty-foot-long minke whale swimming uncharacteristically close to shore in one of the small bays along the west coast of Vancouver Island. Very suddenly, the minke stops swimming, and lolls on the surface in the very center of the bay. It does not move for several minutes. Next, a pod of killer whales is seen rounding the point, heading directly toward the minke whale. The males with their distinctive six-foot-long dorsal fins advance toward the docile minke; the females stay back. One killer grabs the lower jaw of the minke, while another pushes its body tight up against the minke's blowhole. Together, the two killer whales push and pull the non-struggling minke underwater. It drowns. Now the females advance. They grab hold of the minke's enormous tongue and rip it out of the whale's mouth. All together, the killer whales pull the corpse underwater again. Several minutes elapse. Eventually the minke's body floats to the surface again. The killer whales have already left the scene, leisurely swimming back out into the middle of the sound. When the human observers motor toward the minke, they notice a very curious thing. There is a disc floating in the sea. It is the minke's dorsal fin, severed cleanly from the body in one four-foot-wide bite. And more curious still, the minke whale has been skinned, flayed from end to end. Stranger still, there is no sign whatsoever that the whales have even fed on the body of the minke. Two months later, these same observers chance upon another minke whale carcass. It too has been flayed cleanly, tongue eaten, dorsal fin severed, body otherwise unmolested.

It becomes so very easy to assign motive and even meaning to such an improbable event. Was this a case of food gathering, or perhaps a premeditated execution? Or there is a hypothesis au courant among whale scientists that speculates that certain species of cetacean, including the killer whale, are able to stun their prey with sound, and from a great distance as well. One whale's interspecies music becomes another whale's kiss of death. I once witnessed a pod of pseudorcas swim noisily through a tropical sea while all the sailfish in the immediate vicinity surfaced with their sails stiffly erect. The sailfish were so dazed that they allowed us to motor right up to them without moving. There were so many of them lying on the surface that the ocean looked as if it had suddenly become host for a sailboat race. Perhaps that minke whale had been similarly stunned, unable or unwilling to escape its fate. The scientific observers at the scene commented that the tongue, and perhaps the skin as well, offered the killers a change of diet from their usual fare of salmon. After all, there are many vivid accounts of killer whales feasting on the tongues of harpooned great whales out on the high seas. That, of course, is true. But it is just as true that orcas are very often seen in the company of minke whales in the area around Buddy's Cove. Sometimes companions, sometimes enemies. These orcas are beginning to sound as fickle as human beings.

Thomas Kuhn, in his landmark book *The Structure of Scientific Revolutions*, comments that there is no real way to absolutely verify any scientific theory. Part of the problem lies in the unavoidable fact that we are only imperfect humans, creatures limited by our observational sensory acumen as well as by our historical bias. For example, Newton's description of celestial mechanics was once considered an irreproachable law of nature. Then, suddenly, it was supplanted by Einstein's theory. Now, at this point in time, we "believe" in Einstein, yet we still utilize Newton's construct. Why? Because Newton's "law" better fits our limited human perspective. Our daily lives do not move near the speed of light; they do not fit into Einstein's mold unless we expand our view to include the cosmic big picture. Newton works just fine in explaining why an eight ball moves when it is hit by the cue ball. Another part of the problem lies in the fact that science itself employs a kind of built-in tunnel vision. Thus a chemist will call helium a molecule, while a physicist calls it an atom. Jell-O is not always Jell-O to a scientist. Yet neither the physicist nor the chemist is incorrect. Ultimately, no discipline can ever see all parts of the picture. At best, we are all poets. Perhaps we would understand better with a little less discipline.

In the case of the vanquished minke whale, my young ethologist friend, J.S., tended to agree with the conclusions voiced by the observers at the scene who were his peers. They spoke the same language. But when I described the exact same chain of events to a theater director, he believed that the execution was more a ceremonial sacrifice than a gourmet meal. For one thing, there was an uncanny synchronicity between the role of the minke and that of the killer whales. They may have been performing *together.* Such an explanation offers one plausible answer to the bizarre lack of frenzy surrounding the event. In its more poetic moments, social biology seems to hint to us that even the most primitive of animals feel emotion, and perform ceremony. In the case of the flayed minke whale, there is a startling analog to the ritual sacrifices of the Aztec. At Tenochtitlan, a maiden was sacrificed at the altar. Her heart was torn from her body and eaten by a priest who commenced to skin her. He then donned the skin while performing a ritual dance.

When I asked the ethologist his impression of the theater director, his immediate reply was a wide grin, and the comment, "Ah, but he doesn't know whales." The theater director said of the ethologist, "Ah, but he doesn't know ceremony." And never the twain shall meet.

The question remains, however, did the observers witness an orca ceremony? Many scientists now agree that the orcas possess a language of sorts. That is, the individual pods vocalize in such distinct ways from other pods that each group of pod vocalizations is called a "dialect." Even pods in the same area utilize distinct dialects. This implies that the social structure of a pod has a history going back as many years as is necessary to develop a distinct dialect.

More so, there are catalogs of orca vocalizations that identify literally hundreds of different kinds of sounds, combinations of sounds, and sounds correlative with specific activity. There are scientists in the Puget Sound area who are beginning to understand the orca-ese of their local whales. But they cannot make heads or tails of the vocalizations of the whales up in the Buddy's Cove area. At this point in time, there are very few students of orca vocalizations who do not employ a methodology based as heavily on linguistics as on animal behavior. In the case of orca language, there are many scientists who are beginning to use such terms as "dialogue," "conversation," and "song." This is the language of the theater.

In the case of humans, the acquisition of language gave birth to an oral tradition, to the comprehension of history, the development of religion. In British Columbia, among the people who have lived close to the whales for centuries—

the Kwakiutl, the Haida, the Tsimshian—there are those who will unabashedly tell you that orca society is, in fact, a highly evolved civilization.

It is ideas such as this one that give added credence both to the ancient myths and to the alternative explanations of contemporary data. Orca: even the name conjures up a vision of power and nobility. Nobility surely—but religion? Civilization? The local Indians claim that orca is a step above human being. But what does that really mean—the difference between a C-sharp and a D—the difference between humans and angels—or the difference between a species who does not attack human beings even though he can, and another species who "harvests" whales for lipstick?

A recent study concluded that young orcas are regularly left in the care of an aging bull or other elders for a set period of time each day. Eventually someone might take the step to call this relationship a school. But what is the nature of the lessons? How to catch salmon? Almost certainly. How about a course in pre-outboard motor echolocation? Implausible? Perhaps. A conjectural statement to be rejected outright? Not exactly. Instead, let us keep the channels open and learn where the orca/human bond leads us.

By now, I seem to be speculating about ideas that may elicit the kind of response usually reserved for psychokinesis or the dolphin/Dogon/UFO connection. If that is so, then we are also led right back to Thomas Kuhn's analysis of scientific tunnel vision—hooked by the belt to the anthropocentric theories and opinions that guide current views about animals. And all the while, there swim the orcas—large-brained and sensitive beings, creatures who vocalize elegantly figured songs to one another, creatures predisposed to upsetting all our wobbly ideas of what does and what does not constitute an "animal," creatures on the verge of guiding us to the edge of a major zoological paradigm shift.

The orca may not exactly be a common animal, but they are known to inhabit most of the world's oceans. Pods of orca are sometimes sighted from the skyscrapers along the Seattle waterfront. Yet, in another sense, they are a mystery. Here is a creature of intellect upon the earth, probably more like an extraterrestrial than anything else that we have to experience. At this juncture in the budding relationship, we humans must exercise utmost care and responsibility in allowing the bond to develop. We must not only try to "know" the orca as an ethologist might, but also step decisively down from our observationalist's pedestal and "get to know" the orca as we get to know a new neighbor. Yet we must never forget that this neighbor is someone outside our own philosophical and environmental context. The orca race is not to be judged as if it were some exotic human culture swimming around in a black-and-white fish suit.

Rather, we meet orca, the killer whale, *Orcinus orca,* the largest member of the dolphin family of cetaceans. They were originally named killer whales by the old-time whalers because they were so often seen attacking and eating the large baleen whales. They also feed on seals, sea lions, walruses, and any of several kinds of fish.

In Greenland, the Eskimos shape their kayaks to look like killer whales. This reverse camouflage strikes fear into the hearts of seals, who immediately climb out of the water to be clubbed to death. It has also been written that orcas can corral and kill a hundred dolphins during an afternoon's forage without actually feeding on more than a few.

Yet I have willingly and without fear entered the waters of Buddy's Cove many times to swim and cavort and sing with the orcas.

## III. The Bambi Syndrome

Interspecies music expresses the clear and simple example of humans communicating with other species of animals through the universal language of music. Like any music, interspecies music communicates the energy exchange of harmony. Like any successful harmony, it is sustained as long as the participants co-create in the here and now. What this implies in actual practice is that the human must first acknowledge the other being as his or her equal. In many cases, the human must actually sit with the animal as a student sits with a teacher. And it is at this point of recognition, when we truly meet the animal halfway, that the relationship finally emerges.

Such sensitivity to the native wisdom of an animal demonstrates its practical worth the very first time that a valid communication manifests itself. You instinctively recognize in your heart that the animal has a mind. This realization unsettles you. All the formerly secure boundaries that sheltered yet separated you from the natural order proceed to disintegrate before your eyes. You are sucked into the very center of an operative and very complex world communication network. Here is a unified web of interconnected beings all living and dying upon the face of Gaia. New feelings and concepts erupt into consciousness. You are greeted by "Orenda," the Iroquois spirit being who represents that power in nature that awakens a sense of wonder and vitality. He hands you "Tlogwe," the Kwakiutl gift of special powers. It is the ultimate treasure that nature spirits give to those who have dared to enter their secret realms.

And now that you have arrived, it is no longer an easy trick to return untouched to that human-centered universe from whence you began this journey.

So many cherished beliefs suddenly lose all meaning. What is an "animal"? The English term has always implied a hierarchy—there were animals, and then, well, us. And what about communication? It wasn't exactly words, nor was it thoughts or even emotions. All these concepts no longer seem to describe the basic mechanism of communication. Instead, they are reminiscent of the outmoded but still useful terms in Newton's physical universe. That is, they work, but only for the limited human scale of reality. They are leftovers of the old anthropocentric worldview.

But let us take care not to over-romanticize the interspecies revelation. After all, who can ever know if an orca *feels* the same unbridled excitement, or *thinks* the same insinuated causality, when a human guitar player succeeds at eliciting a slowed-down repetition of a certain stereotypical musical phrase? To the contrary: the key to a continuing communication relationship with orca, or with any animal for that matter, is, in fact, not to *feel,* not to *think.* Just be! Become a creative and patient conversationalist and persevere. That way the relationship may go in any one of several directions all at once, unencumbered by the restrictions of the analytical human sensibility. Now, it is free to enter into realms that none of us has ever even considered. Neither you, nor I, nor orcas.

This reason alone explains the redundancy and shallow insufficiency of the many experiments in interspecies communication that are carried out on captive animals. When the experiments are human-centered, they must produce human results. They tell us absolutely nothing about the wisdom of the animal, and in fact, practically nothing about the animal's ability to communicate either. They succeed merely as indicators of whether or not any animal has the ability to mimic certain human intellectual models, in return for the major necessities of food and companionship. The animal is never a participant, always a specimen. These experiments are always undertaken from the medium of the cage, the house trailer, the concrete pool. As Marshall McLuhan says, the medium is the message.

And success is a dolphin who, after five years in captivity, learns to vocalize like a human two-year-old, a chimpanzee who can sign a number of words in Ameslan but whose psyche is characteristically described as "not much like a chimp anymore." Imagine an extraterrestrial who communicates through a series of very complex but nearly imperceptible face movements. He lands on the earth, ditches his hyperspace vehicle, and is somehow captured and put into a cage for study. After five years, the extraterrestrial, through a monumental and painful effort, and in an attempt to "please" his captors, learns to sign two hundred words. However, he cannot learn to connect the words together, because his culture eschews intellectual communication of ideas. That was something they did thousands of years ago when they were first learning how to manipulate machines. His constant smile says that a picture is worth a thousand words. But the earthlings in their white coats, they just don't understand.

If at first this analogy seems a bit precious, then try to comprehend the fact that a dolphin or a chimp in the wild is a kind of extraterrestrial. Its reality is, for the most part, outside the normal human perception. Dian Fossey, who lived with mountain gorillas in central Africa, remarked that she was initially accepted by the resident gorilla troupe only after she had learned and practiced a basic gorilla vocabulary of complex body movements. One wonders, how might a talented dancer, or an Aikido master, have learned to live and communicate with gorillas?

I am especially fond of Douglas Adams's description of the language of the dolphins.

> The dolphins had long known of the impending destruction of the planet Earth and had made many attempts to alert mankind of the danger; but most of their communications were misinterpreted as amusing attempts to punch footballs or whistle for tidbits, so they eventually gave up and left the Earth by their own means shortly before the Vogons arrived.
>
> The last ever dolphin message was misinterpreted as a surprisingly sophisticated attempt to do a double-backward somersault through a hoop while whistling the "Star Spangled Banner," but in fact, the message was this: "So long, and thanks for all the fish."

According to Adams's *The Hitchhiker's Guide to the Galaxy,* humans are only the third most intelligent earth life form. Dolphins are second, and white mice, "who conduct frighteningly elegant and subtle experiments on man," are first.

Or consider another profoundly humorous anecdote about the tribulations of interspecies communication, this one from Sir Richard Burton's nineteenth-century translation of the Hindu classic *Vikram and the Vampire.*

> In the days of old, men had the art of making birds discourse in human language. The invention is attributed to a great philosopher who split their tongues, and after many generations produced a selected race born with these members split. He altered the shapes of their skulls, by fixing ligatures behind the occiput, which caused the sinciput to protrude, their eyes [to become] prominent, and their brains to master the art of expressing thoughts into words.
>
> But this wonderful discovery, like those of great philosophers generally, had in it a terrible flaw. The birds, beginning to speak, spoke wisely and so well, they told the truth so persistently, they rebuked their brethren of the featherless skins so openly, they flattered them so little, and they counseled them so much, that mankind presently grew tired of hearing them discourse. Thus the art gradually fell into desuetude, and now it is numbered with the things that were.

Some propositions can never be proven or disproved within a logical system. So states Gödel's Theorem, whose major application lies within the world of computer programming: A computer of any given size can only replicate or model a smaller computer. John Lilly, who must be considered the father of modern interspecies communication, argues that the brain is a biocomputer. Thus, an orca whale, with a brain three and a half times the size of ours, could model a human—but not vice versa. This analogy of brain to computer is certainly useful up to a point, but it seems to break down miserably as soon as we try to quantify brain size as if it might define the number of bytes available to any individual brain. It is a classic mistake of humans who think they understand animals.

A smaller brain, such as the one that resides in the fertile head of a seagull, is not precisely a biocomputer. After all, we larger-brained humans can no more

map the brain of a seagull than we can the brain of an orca. The reason is very straightforward, if not also a wee bit tricky: brains "contain" minds and likewise, minds "contain" consciousness. As we attempt to study minds and consciousness, we find them quite unwilling to remain categorizable. Minds are slippery; consciousness is unpredictable. And neither remains very transparent as far as its abilities and functions. In fact, there are many scientists who remain skeptical that a seagull might possess a mind at all, much less a consciousness. On one level, many of these skeptics fall prey to a blindness I choose to call the "Bambi Syndrome." That is, they cannot accept the reality of *animal* consciousness until an animal possessed of *human* consciousness appears on the scene.

At this particular juncture in our natural history, we humans tend to manifest our consciousness through the twin pillars of *thought* and vocal *language*. The theory of the human bicameral mind states that human consciousness was very different as recently as the fifteenth century. Can we not assume an alien, but very sentient, consciousness for seagulls? Perhaps seagull consciousness is based on something besides thought. How do we imagine such a thing? Or stated another way: No human can ever prove or disprove the ability of a seagull to communicate great profundities to another seagull.

Or stated yet another way: What is this word "profundity"?

My imagination turns the poet as I contemplate the realm of seagull consciousness and communication. Perhaps a seagull "speaks" through a subtle variation in its flight patterns as is so vividly described in *Jonathan Livingston Seagull.* It may greet other seagulls differently than it greets an albatross or a piece of kelp, or a kelp forest. All three variants may contain information about water temperature, tide cycles, and tomorrow's wind velocity. A seagull engages a tern in a lively discussion about holophonic obturation (a concept beyond the mapping capability of the human biocomputer, but based on the general idea that a smaller brain possesses less tunnel vision) by slightly dipping its left wing over an eleven-foot wave. What does a seagull say to an orca? It may analyze Darwinian selection by swooping low around the orca's dorsal fin. The orca may agree with every single one of the seagull's points by eating him.

In my personal and sometimes foolish need to comprehend cause and effect, I was once very sure that a seagull said hello to me by shitting squarely on the top of my head.

The question of interspecies communication and animal language becomes increasingly provocative when we turn our attention to the relationship between humans and orcas. It is not so much that an orca is more intelligent than any seagull. We have already shown such a criterion to be a conceptual dead end. Instead, let us simply comprehend that an orca possesses a large brain, and thus, its apperception of both itself and the world is closer to that of our own large-brained consciousness. Here are some of the crucial considerations: First, we humans stand poised at the brink of accepting, and possibly even unraveling, orca language.

Should we consider any ethical constraints in commencing such an endeavor? Do we engage the orcas in a dialogue, or rather, do we focus upon what they are saying to one another? Second, humans and orcas are already relating to one another in a discernably sophisticated fashion within the confines of that well-established human communication mode known as music. Music expresses volumes of information about such concepts as creativity, emotion, culture, psyche, and rhythm. Furthermore, when two creatures generate a true harmony, the very air and water begins to shimmer. On that level, orcas and humans are mindfully recreating a classic experiment in acoustics. Do the orcas possess their own music? Third, the process is never merely a collaboration between orcas and humans. Perhaps more important still is the matter that here are two *individuals*, beyond the confines of species, environment, and history. Maybe only one in every five orcas possesses a musical imagination. Fourth, and the key to understanding why I choose orcas over seagulls, is the fact that the music itself is jazz. I, for one, have indeed tried to make music with seagulls. Yet I have never received a reply that can remotely hold my musician's attention the way a lead orca can. However, I can easily envision a human glider pilot fathoming the rich lyrical flight poetry of a seagull. They both speak the same language. Finally, orca language is frequency modulated and audible. A musician translates this jargon as a melody that he can hear. Compare this with the vocalizations of almost all other dolphin species, whose calls exist mostly above the limitations of the human ear. Thus, this musician prefers orcas to other dolphins.

So many possibilities, so many intertwined levels of meaning all curling inward toward the center like the petals of a rose blossom. Yet, as the construct grows weightier and weightier, suddenly all the possible answers evaporate away into nothing at all.

Ultimately the experience with an animal is going to rise up in front of you as intimately as a lover and quite devoid of intellectual content. How can you deign to say that "this means that" while experiencing a connection bridged across inconceivable gulfs of imagination and perception? When we try, we must fail. And if we still need some all-encompassing rationale in order to accept the cogency of the human/orca bond, then let us ponder the koan of Compton's principle of reciprocal action: There is no known phenomenon in which one subject influences another without itself being influenced at the same time.

What a pity that I am unable to describe the interspecies relationship utilizing the language of the Iroquois, the Tibetan, or the Hopi (where there is no past or future tense). How much more informative and to the point it would then appear. The English language itself lacks the finesse of these other, non-technological, earth-centered tongues. To paraphrase Gödel: English emerges as one more example of a system unable to prove or disprove all propositions. It is too small a computer to map the much larger unity of interspecies communication. However, as we English-speaking humans have come to comprehend the concept of the profound, so is this a profound connection. It turns all our cherished ideas about the earth, the human race, and especially our own individual minds inside out and upside down. It is like describing a sound with just your toes, like describing a fuchsia to a blind man. Mumon, the Zen poet, said it so very well:

> Has a dog Buddha-nature?
> This is the most serious question of all.
> If you say yes or no,
> You lose your own Buddha-nature.

## IV. Listening and Playing

It is well past midnight, probably as late as 2 A.M. The air is rich and thick with a fog that squirms inside your clothing, slithers inside your mouth with each passing breath. When this kind of late-night fog settles over the Pacific Northwest, the environment takes on the classic mood of mystery. Nothing moves; nothing makes a sound. It is the stuff of owls and howling wolves, monsters and headless horsemen. I am feeling more than a bit like a horseman myself—comfortably nestled inside the cockpit of a fiberglass kayak, all wrapped up in matching yellow raincoat, rain pants, and knee-high rubber boots. They are all doing an excellent job of keeping my body warm and dry.

I am sitting inside a sleek kayak playing an electric guitar through a battery-powered sound system. The sound is transmitted directly into the water through a saucer-shaped underwater speaker. The music may travel as far as three miles underwater. Above the water you've got to listen hard just to hear it at all. It sounds like some kind of nether reggae dream tune.

It can get dangerous out here in the dark, especially if the wind comes up or the current moves too strongly. I am sitting two hundred yards from shore at the outer lip of Buddy's Cove. The only practical method for determining position is to keep one ear firmly focused on the gentle waves that lap against the beach. It is essential. This water never gets warmer than forty-five degrees Fahrenheit. Sitting as I am, two hundred yards from shore, the bottom drops steeply down to four hundred feet. Out in the middle of the strait, a mile farther out, the water is eighteen hundred feet deep. In general, this is a sharp up and down country, dug out by two-mile-thick glaciers ten thousand years ago. The strait itself is about three miles wide and a hundred miles long.

The fog has obliterated the light of the full moon as thoroughly as if I were inside a cave. But the water itself is alive with the light of millions of bioluminescent organisms who turn themselves on and off with the regularity of a neon city. Dip the paddle into the water and the swirl of the blade stirs up the creatures to new heights of display. The part of the paddle that is below the surface looks exactly like the light sabers from *Star Wars*. It is a striking green-white color. It all makes me feel as if I'm drifting upside down—the sky above is black and water-filled; the water below is jam-packed with stars.

The kayak paddle serves as a brake, to prevent the imperceptible but persistent current from carrying me out of the bay. In this fog, the resultant loss of position could signal a disaster. In this case, the bioluminescence is a godsend. It may be too dark to see if the kayak bow is cutting through the water. But when it does, I can always tell, because the organisms themselves always light up the direction.

A hydrophone dangles twenty feet below the surface to starboard. The much heavier speaker eerily glows ten feet below the surface to port. Tonight it looks more like a real flying saucer than anything that I've ever seen in an illustration. I am listening to the underwater environment through a pair of headphones. Every minute or so I pull them off my head and cock an ear toward the direction of the shore. Good, it's still there.

Ever since a log barge chugged up the strait a half hour ago, I have not been able to hear anything but the incessant crackle of the current flowing across both the kelp and the sides of the kayak. Just a few minutes before, I heard a surge of something below. In fact, it was a school of shrimp passing directly beneath the kayak. Each individual shrimp generates a single "click" with the beat of its tail. All together, a school of ten thousand shrimp sounds something like a radio between stations. Herring feed on the shrimp. Salmon feed on the herring. Orca, the killer whale, feeds on salmon.

Every thirty seconds or so I punctuate the crackling ocean with a short, three-note guitar burst: D-C-D, D-C-D, D-C-D; three times, then silence again. I'm not trying to impress the orca with anything fancy, no grandstanding flourishes of virtuosity. The skill comes in properly using a glass bottleneck over my little finger. That way I can slide the glass along the string, one note to the next. It gives the phrasing a fluid unity, rather than having it sound like three individual notes. I begin by bending the D note slightly sharp, striking the string and then quickly dropping the bend down to the D, dropping the bottleneck onto the string and sliding it down to C. It is important that the time taken between the descending D to C be at least twice as long as the return back to D again. In effect, I am approximating one of the more common phrases from the language of this particular pod of orcas. Maybe it means "hello." One sarcastic friend has suggested that it is most certainly an orca swear word.

Tonight I am playing a brand-new soprano electric guitar. This instrument reaches nearly a full octave higher than a normal-sized guitar. The frequency range of a musical instrument becomes critical when trying to match the orca as he or she vocalizes into the very high-pitched region of squeaks and squeals. Furthermore, this wonderful little guitar is only two feet long. It is absolutely perfect for playing in the cramped quarters of a kayak. I have built a waterproof guitar case right onto the front deck of the kayak.

I am not alone out here tonight. Richard Ferraro, a research scientist who himself has become mesmerized by the acoustic interaction between humans and free-swimming orcas, is recording the entire event from a kayak two hundred yards to the south of me. My wife, Katy, has tethered her kayak to Richard's. The two of them are listening to the underwater environment through a multi-headphone system that Richard has installed inside his boat. I have separated my kayak from them to permit a better-balanced sound recording. Until the orcas actually venture close, I am usually transmitting quite loudly. The guitar sound can clutter up the taped result unless the speaker is properly separated from the recording hydrophone. Then, when the orcas come close, I immediately turn the amplifier way down to match the orcas' volume. Richard and Katy re-maneuver into the best recording position, and so the result sounds balanced and crystal clear. "Hey, Jimmy, let's take a short break!" shouts Richard. "Pull up your wires and paddle over here. We've got some Wheat Thins, cheese, and wine."

I unzip the neoprene spray skirt, constructed to keep the water out of the kayak's cockpit, and reach underneath it to turn off the switch that controls power to the sound system. Next, I unplug the little guitar, hold it in my lap while I rip open the Velcro seals of its fiberglass case. Inside it goes, then sealed once, twice, three times. Next, I reach under the seat, pick up the cable spool, and roll in the hydrophone. The tiny, pencil-thin hydrophone is securely buttoned to a special cleat built into the left gunwale of the boat. Finally, in comes the heavy speaker. This is placed inside a fiberglass dish located on the deck, just in back of the cockpit. The entire process of setup and breakdown takes about two minutes.

Using all this specialized gear, over the years I have learned a few things about how to best communicate with an orca. The first rule is: Don't ever chase them. Even when they come very close to echolocate on the speaker, satisfy their initial curiosity, and then begin to swim away. Even then, let them go. Anything else can only be construed as harassment. If they are interested in the relationship, then they will certainly come back on their own. And if they don't? It is because they have something else that interests them more. That is the major difference between what we do out here in the fog, and what the institutionalized researchers of interspecies communication do at the oceanarium. A wild animal is an equal partner in establishing the process. Ultimately, it makes for a much more communicative relationship.

Second, we try to play from the exact same spot, every single night. That way, when the whales eventually make their way back to the spot where they heard that music yesterday, or last week, it may happen again. We thus establish a continuity of place. We've gone so far as to give this music-making locale a formal name: the *svaha,* from the Sanskrit, meaning "a space within a space."

Third, schedule the communication sessions for the same time of night. That way it establishes a continuity in time. It is best to play at that time whether there are orcas present or not. And don't rush off to the kayak every time the orcas cruise through the area. The best time is after the sun goes down. After dark, nearly all the commercial fishing boats, pleasure boats, and whale-watching boats head back to their moorings. The ocean itself seems quieter. Likewise, the orcas seem much more enthusiastic. Maybe it has something to do with their scheduled rest or play period. I don't know.

And finally, play as few notes as is absolutely possible to keep the exchange alive and interactive. In other words, be a good *listener.* I have spent much time analyzing the tapes made during the five years of my ongoing relationship. I have learned that, in the beginning, I would get just too excited every time the orcas responded to my sounds. And every single time, it signaled the beginning of the end of that session. It has been three years since I first heard the orcas singing reggae with my guitar rhythms. It has been two years since I sat in front of a campfire swapping stories with the filmmaker and the ethologist. Last year I spent a week up in Buddy's Cove being filmed for American TV—in the water, while the orcas drew close and slapped their tails for all the world to see.

I have spent many hours since then learning to incorporate cool-headedness, slow-handedness, and a bit more seductiveness into my guitar playing. I spent weeks listening to Japanese koto music and Miles Davis solos, as exercises in how to get more from less—the minimalist's approach to interspecies improvisation. Over time, first listening to all the guitar/orca sessions and then playing along with them, I have learned to utilize each individual note as a reflection of the conversational mood of a friendly orca. What I had first heard as a straightforward D-C-D progression, I now hear as any number of uniquely varying calls. That supposed D note, is, in fact, almost never *D,* but rather twenty or more subtle shadings of pitch from C-sharp up to E-flat.

Communicating with orcas has turned out to be no different than learning any foreign language. At first, everything sounds like one long unbroken string of indistinguishable syllables. Then with time, and much practice, individual words, phrases, and inflections begin to stand out from the monotone. If there is a difference between learning orca and, say, Nahuatl, then it is twofold. First, there is no dictionary yet available to point out the meanings of certain words. Second, those to whom I converse are almost never in sight, so they cannot or will not take the time to point to the objects that they are speaking about. And so, for the most part, after five years of study, I do not yet have any idea how those words and phrases translate into English. Maybe the entire concept of a Rosetta stone is irrelevant. Maybe I'll learn something tonight.

The kayak slices through the glassy water like a knife. It takes less than ten strokes to traverse the two hundred yards that separate Katy and Richard from me. Yet despite the speed, my boat still feels quite a bit slower than normal tonight. The dead weight of a fifty-pound battery, stowed in the kayak's bulkhead, would tend to slow anyone down.

Katy is quietly humming an old jug band tune: "Ain't ya glad, ain't ya glad, that cars can't talk." But in general, there's not much of an attempt at making small talk out here tonight. The dense fog, the bioluminescence, the lapping of the waves on the shore, the weighty anticipation of the orcas who may eventually come around to interact with us—what in the world can we talk *about*? We sit here humbled in the face of the environment, no more than three human energy fields, floating around in the altered perceptions of a dream. No, that's not right; it is not exactly a dream. We're all too much awake. Maybe it is just that the props of the situation are so overwhelmingly dreamlike. These props do have a power all to themselves. Richard comments that he's glad he has a tape recorder to document all the underwater sounds. He does not see how he could ever trust his powers of recollection.

And then, very suddenly, both Katy and Richard let out a short excited "Hah!" in unison, and cup their palms over their headphones. They can hear the sounds of orcas vocalizing underwater. "Hah! Did you hear that? They did it again." Without headphones, I hear nothing. But I do have a strong idea what it must sound like—a cross between a violin and an elephant—the vocal call of the orca. I perk an ear to listen for the characteristic sound of a whale surfacing for breath. Nothing. That means that the orcas must still be over a mile away. I wait patiently and continue to sip a glass of Cabernet, breaking off strips of the stringy provolone cheese, and eating it with slices of orange. Katy comments that the sounds are getting louder. The whales must be headed in our direction.

Now I hear them blowing on the surface: *Pooo! Pooo! Pooo!* They sound about a half mile away. Richard slips his headphones off his ears and asks me to paddle my kayak back to where I'd been before. I down the last of the wine in my glass, stow it inside the pocket of the jacket. I dip the paddle into the water, and paddle ten strokes toward the whales. Next, stow the paddle, unzip and unwind all the kayak's assorted built-in gear, drop the hydrophone over one side, the speaker over the other. I put on the headphones and give a listen: *Weeeeeeeeeeyyyyyyyy-ooooooo-oosh!* There is an orca vocalizing underwater. Sounds about four hundred yards away.

The very first guitar sound is critical. It needs to be timed to commence just after an orca finishes its phrase. This helps establish the general format of *dialogue.* Furthermore, my own phrase should start off with the orca's *last* note, to establish a continuity of pitch. Now an orca vocalizes again: the legato phrase E down to C-sharp. In answer, I bend the C-sharp note up about a quarter-tone, strike the string, quickly drop the pitch down onto that C-sharp, and then slide the bottleneck up the string ever so slowly until it lands squarely on the E, where I wiggle the string back and forth to generate a vibrato. At this early stage in the conversation, I am not trying to match the orca's sound precisely. That will come later, after the form of the conversation is better established. For now, I want to create a human's musical *interpretation* of the call. That way, if the orca's phrase is some kind of identifying signal, then I can be sure that my signal in similar style will also identify me as a human wishing to communicate. The orca will readily hear that my call has some subtly unique characteristics of its own. That these *characteristics,* for example the vibrato, are based upon *musical* parameters, gives it firm footing in a very ancient, universal form of communication. I consider music to be the highest expression of communicating humans.

A heavy, ten-second silence elapses at the end of my phrase. Then, the orca vocalizes its own phrase as it did before; this time it repeats the phrase twice. I repeat my phrase again, now twice. The orca picks it up just as I land on the C-sharp; but this time it starts on E-flat, drops to C-sharp, wiggles the note back and forth, and then slowly rises up and up until the sound goes beyond my hearing range. Silence. This time, I answer by sliding up the B string from the high octave B up to E, change over to the E string, and continue to slide up right to the bridge of the guitar. Still, I retain the exact meter of the orca's phrase. However, it does not work as well as I had hoped. Somewhere in the middle of my phrase, the orca cuts in and phrases E to C-sharp again. It breaks the form of dialogue.

There is only one orca vocalizing. That is a good sign. I play my own first phrase: C-sharp to E again. The orca repeats it. That is a good sign. Slowly, ever so slowly, the form of a precise dialogue is taking shape. It demonstrates a situation of truly listening to each other's voice.

Now I can hear the whales blowing on the surface even with my headphones on. *Pooo! Poooooo! Pooo!* The three of them are swimming directly between Richard, Katy, and me. The lead orca is vocalizing very loudly, so loudly that I have to turn down my headphones. There he is! I cannot actually see him. But what I do see is a long bioluminescent tube formed by his body as he swims below my boat. It is almost certainly a male, probably over twenty feet long.

Suddenly, I hear the sounds of more orcas calling, but from very far away. Now, the lead orca vocalizes once again. This time it is a very different quality of sound, somehow richer and more varied than anything that has come already. I remain silent. Better to say nothing than to intrude in what may be communication between the pods.

Furthermore, I feel that I've heard this sound before, and in fact, I once sat in front of a computer screen analyzing its component parts. In the earlier rendition, the orca was saying three things at the same time. First, it was echolocating on the speaker. Second, it was matching a guitar phrase. However, the third overtone of the matched phrase seemed to have a life all its own. On the screen, the voice pattern looked more like the loopy, curvilinear letters of the Arabic alphabet than the characteristic parabolic swoop of a "normal" vocalization. It was an overtone possessed of unique and quite complex bits of information.

And this time, when the orca vocalizes the three-in-one sound, it is answered by the orcas off in the distance vocalizing once again. Then, as if in response to the distant call, our three orcas quickly swim off, heading north in the direction where we expect the other orcas to be waiting.

Silence. The orcas are gone. I let out a deep sigh, pull in the sound gear, and slowly paddle back to Richard and Katy. The three of us sit quietly in the middle of the cove chewing on Wheat Thins, watching the lights play off our dangling kayak paddles. Katy shines a flashlight in the direction of the breaking waves. The shore is nowhere in sight. The fog is just too thick. Richard quietly comments that he is beginning to hear the orcas again. Katy adds that she can hear them blowing. From the number of blows in quick succession, she feels there may be more of them. I paddle off into the dark to set up the sound system once again. *Poooo! Poooo! Poooo! Poooo! Pooo!* It's hard to tell for sure, but there may be as many as seven of them this time. Richard shouts out across the cove, "Hey Jimmy, it sounds as if we might have the beginnings of a real party!"

Once again the orcas enter Buddy's Cove. And once again, there is only a single orca vocalizing with the guitar. There may be no way to prove that, but nevertheless, I seem to sense it. The orcas head directly for our kayaks. I can hear them echolocating strongly into the hydrophone, a sure sign that they are within a hundred yards of me. With headphones on, the clicks of their sonar sound like nothing so much as a barrage of machine gun shelling. *Tckatckatclcatcka.* Now they are very close. I hear them echolocating even with my headphones off. The water beneath the boat lights up as first one and then several other whales swim just in front and then underneath. It looks as if the ocean is on fire. I can do nothing but sit there and watch.

Then I play the same riff: C-sharp to E. Played once, and suddenly all the whales begin vocalizing loudly—and all at the same time. At first I try to pick out one particular call. Unfortunately, my rendition is lost in the dense cacophony. Then, silence. I pick out a full A-major chord. No reason. Maybe I'm following my fingers. Still silence. I hit the chord four times in tight rhythm and move the chord up to a D-major. Four times and then back again to A. Silence. Then, from very far off, I hear the vocal calls of still more orcas. And as if on cue, the lead orca answers them. Within another ten seconds, all the orcas very quickly leave the scene, heading north. They are all vocalizing as they head off into the night.

I pull off my headphones, and notice, by the shining of a flashlight, that Richard and Katy are sitting no more than fifty yards from me. As they paddle close, Richard repeats over and over again: "I had no idea, I had no idea," in a voice heavy with exhilaration. He wonders out loud if it's time to seriously consider heading back to shore. "Look," he states, "the fog is getting thicker. Katy pulls up beside my kayak, and breathlessly tells the story of how all the orcas had surfaced and blown right next to their boats.

For a long moment she had wondered what those huge animals with their monolithic fins had in mind. It is one thing to consider the orca a highly evolved spiritual being all full of sweetness and light while planning a field trip from the civilized security of one's dining room table. But it is quite another thing to have that same creature blow hot air into your face through the thick fog of a 2 A.M. wilderness.

Just as I am about to agree with them that we should head for shore, the nearly synchronized blowing of several orcas makes us change our minds. Both Richard and Katy let out little giggles. The orcas are demanding our full attention. Once again, Richard and Katy paddle off into the darkness.

This time, it seems as if there may be as many as nine whales, if their blows are any indication. They sound as if they're all floating right on the surface, and no more than a hundred yards away. Just waiting for something to happen. I tentatively let loose with a single long-sustaining E note. Silence. More blowing. Suddenly, a single orca calls out the obbligato E to C-sharp and then a slow glissando back to E again. He or she repeats it. It almost seems as if the whales had left the area to discuss the situation among themselves, and have now regrouped around the kayaks with a definite plan in mind. Everything about this relaxed encounter seems more organized.

Back and forth goes the orca/guitar dialogue. Only a single orca is vocalizing. At one point I initiate three identically spaced E notes in a row. The orca answers the phrase with two E notes in identical rhythm. I answer with two E notes. The orca responds with a single E note. I repeat the single note. Now there is a deep silence. For ten seconds the affirmation of the zero hangs over Buddy's Cove with an entire life of its own. From two hundred yards away, I hear Richard let out a loud chortle. Good, he too has caught the count from three to two to one to silence. I recall the dolphin trainer who once informed me that sometimes his animals can count as high as twenty-five.

Do I respond? That zero must signify something. Is it possible that the orca and I just agreed that we have nothing more to talk about tonight? The silence settles deeper. Now the orca vocalizes again. I answer. And so the magic moment is neutralized. Suddenly all the orcas break up into smaller groups and then exit Buddy's Cove. Richard and Katy paddle over to me. "Did you hear it?" "Yes, I heard it. But did you hear it?" "Yes, I heard it too." Richard whispers that he's gotten all three encounters on tape. Good! Katy reminds us that we should all head into shore. Tomorrow, we have to wake up early in order to break camp and then paddle the ten miles to the parking lot up the coast.

We are rudely awakened at 8 A.M. by Kelly, the camp leader of this expedition. "Get up! Get up!" he shouts. "Look out on the water!" Bleary-eyed, Katy and I pull ourselves out of the snug sleeping bags, zip open the tent, and step outside. It is a fine, sunny day, the warmest morning of the entire trip. We step through the forest, to the edge, where a five-foot cliff separates the march of the trees from the pebbly beach. And there, just fifty feet from shore, swim a group of four orcas. Back and forth they go, now closer, now a bit farther out. They are acting more interested in arousing our attention than I have ever seen an orca. One animal seems especially gregarious. She breaches once, twice, her entire body lifted completely out of the water. Four kayakers from our group have quickly dragged their boats down to the water and are already paddling out. The orcas lift their heads completely out of the water and actually peer directly into each human's face. Katy and I burst out laughing. Could that be what they were doing last night?

Within a few minutes the orcas head out into the middle of the strait and are gone. I am simply too sleepy and uncoordinated to hurry down to my own boat, pack in all the electronics, and paddle after them. That is no way to work with orcas anyway. Better to leave their morning greeting and departure as a finale to great interaction. There will always be next year.

# 10

# *The Human/Dolphin Community*

THE DOLPHINS ARE STUDYING PEOPLE. At Monkey Mia, along the west coast of Australia, bottlenose dolphins swim right into the shallows of a beach, and there survey all the humans who wade out to meet them. They seem to prefer the company of children over adults, women over men. Similar situations occur off the coast of Brazil and Hawaii with spinner dolphins, the Bahamas with spotted dolphins, and Wales and Cornwall also with bottlenose.

An analogous experience exists at Buddy's Cove. When we play music in the vicinity of an orca pod, the whales often make a credible effort to plumb the depths of both communication and communicator. Like many human musicians, they seem to prefer to work at night. By sharing time and space through a song, we openly begin to participate in the creative process of each other's lives. That is one viable reason for making a commitment to the concept of human/dolphin community—because it is already happening.

Any such formative community must be based upon a pact of interspecies comradeship, with dolphins as free-swimming partners in the developing process. Thus, we eschew the heavy-handed human preoccupation of control through capture and captivity. What possible purpose can it serve to render the very social dolphin down to a Specimen, a cramped body, a brain to be filled to the brim with human projections? As Wade Doak has written:

> Benign methods of field research into cetaceans depend upon the emergence of a cultural climate among humans in which trust and mutual acceptance are widespread. For this reason, non-benign treatment of dolphins, such as their hunting and capture for exhibition or tagging, are in total conflict with this ethos and forestall the growth of scientific knowledge of these creatures.

How much better might we fare if our communicating partners are allowed to come and go as *they* please. The human/dolphin community grants equal time to the dolphins to teach us, and at their own momentum.

The human/dolphin community may prove a major evolutionary step for contemporary human culture—for humans to meet and mingle with another earthly creature from within the intensive framework of community. The concept seems a prime example of what co-evolution means: a tenuous first step away from our human role of exploiter to the new role of treating the planet as home and neighborhood. It is a modern realignment to the unifying wisdom of totem, a commitment by a group of humans to address the dream of one-world community. But such terms are often interpreted smugly; "one-world community" often seems a buzzword in its airy naiveté. So let us first attempt to ground the concept in the pragmatism of ecology.

Barry Commoner's first law of ecology states that everything is connected to everything else. In that sense, the one-world community exists whether or not we decide to recognize it as such. The verdict on the human use of the environment is all in. It seems urgent that we, both as individuals and as a species, make the commitment to discover some working method to better tune into this law of ecology. But who among us is able to provide the model? The artists? The zoologists? In fact, there are many such models from any number of disciplines. I choose to focus on a few from the worlds of biochemistry and biophysics.

On the building block level of genes, there exists far more similarity than difference between snails, algae, lizards, and human beings. Every living thing on earth shares its birthplace in the primeval soup of 3 billion years ago. Add a bit more hydrogen at one point along the DNA double helix and you may someday blueprint a starfish. Alter a nucleotide there and presto!—an orangutan. Of course, I stand guilty of oversimplifying what is an unimaginably complex process spread out over billions of years. However, the point is still made: All living creatures are remarkably related on that basic level of genes. We are all the same stuff. Geneticist Richard Dawkins pushes the intent of the genes nearly into the realm of consciousness. He postulates that the process of evolution might rightly be described as the struggle of genes, who seek the optimum "survival machine" to best meet the challenge of a particular epoch. As such, you and I, outfitted with large brains, an erect posture, and prehensile thumbs, might well be categorized as the latest trend in an ever-changing line of fashion designer genes. (Pun upon nonpun.)

Such a blatantly reductionist conclusion fails to tell us anything about all the myriad processes in which both the genes and the survival machines are so inextricably immersed. One cannot predict the United States economy by examining one brick from a Wall Street skyscraper. Or for a better example, one that relates cogently to our subject of community: All the genes that make up all the individual life forms of the salt marsh tell us nothing whatsoever about the interdependent community of that marsh. In fact, there are those students of holism who would describe the community of a salt marsh as an organism unto itself—the next exponential level of survival machines.

The Gaia hypothesis postulates that the earth itself is a self-regulating organism. By some mechanism, not yet clearly understood, all the living beings and communities of beings have been living and breathing cooperatively since the very inception of life on earth. This ongoing process regulates the dynamic stability of both air chemistry and temperature. But in reality, what does it mean that all life *cooperates* to ensure the stability of the biosphere? Such a process strongly implies that some vast global communication network exists, and that it functions across both space and time. If such a "network" is fundamental to life, then we humans must possess it as well.

Do you, the reader, experience "it" at this very moment? And how shall we describe "it"—a bioradio? A field? Perhaps it assumes different forms in different beings. Hopefully, it will manifest in human beings as an ethical constraint against building more bombs, deforesting more jungle. Ethics as one possible large-brained manifestation of Gaia.

Throughout this book I have attempted to describe a communication network that I, personally, have experienced many times while communing with various species of animals. Other writings, such as *The Secret Life of Plants,* suggest that some researchers have experienced a similar phenomenon in the presence of plants. Although I take responsibility for my own described experiences—for example, the buffalo's social aura or the bizarre images conjured up in the presence of gray whales—still I find myself lacking the language skills to describe this natural wisdom without sinking into an inevitable soft-core romanticism that can only mar its import. I defer to Lao-tzu:

> The Tao that can be explained is not the real Tao.

Biochemist Rupert Sheldrake described a natural phenomenon that may bring this natural communication network further out into the light. Although his controversial hypothesis, called formative causation, and best described in *The New Science of Life,* remains a matter of speculation—not devoid of its own internal inconsistencies—still, it offers a highly provocative explanation of all that is most relevant here.

According to Sheldrake:

> The orthodox approach to biology is given by the mechanistic theory of life: living organisms are regarded as physico-chemical machines, and all the phenomena of life are considered to be explicable, in principle, in terms of physics and chemistry.

Unfortunately, this established paradigm does not account for the fact that organisms possess properties that cannot be fully understood in terms of their parts in isolation from each other. The whole is always more than the sum of its parts. Sheldrake found the anomalies especially acute in explaining morphogenesis: the way that organisms take their own form. New structures are always appearing that cannot be explained in terms of the structures present in the embryo. Many recent experiments imply, instead, that the characteristic forms and even behavior of physical, chemical, and biological systems are determined by invisible organizing fields that act across space and time. These Sheldrake calls morphogenetic fields. Sheldrake further hypothesizes that these fields are without mass, and even without energy as we now describe the term. In fact, morphogenetic fields have been previously unknown to science, although they have been at least implied by such bold theoreticians as Paul Weiss and Wilhelm Reich. In many ways, morphogenetic fields seem more the province of shamans and mystics, people like Jackson Jacobs and Frank Robson, than of biological science. That is precisely why Sheldrake's hypothesis may prove to be so important.

Sheldrake bases his hypothesis on the results of several unrelated experiments. Space permits me to mention only one. In 1920, working at Harvard, psychologist William McDougall began a series of experiments designed to systematically train related generations of rats to learn how to escape from a water maze by swimming to an unlighted gangway. If the rats swam toward a lighted gangway, they would receive an electric shock. McDougall measured the rate of learning by counting the number of errors they made before they finally figured out how to swim to the unlighted gangway. Over time, he found that although the task never changed, later generations of rats were definitely learning the task more quickly than the generations that had preceded them.

This effect persisted even after McDougall bred only the slowest-learning rats from any one generation, so that the learning rates could not be figured as a simple result of heredity. Yet over twenty-two generations, he documented a tenfold increase in the rate of learning. Later, separate studies in Scotland and Australia were designed to replicate the original experiment. Significantly, and in both cases, the very first generation of rats learned just about as quickly as McDougall's last generation. It was as if the rats were somehow able to communicate the correct watercourse way through some mechanism that cut across both space and time.

According to Sheldrake, the first generation of rats established the morphogenetic field for the specific behavior being learned. That field then "guided' the subsequent generations. Each generation reinforced or prolonged the power of the field. Eventually, what Sheldrake has called a *morphic resonance* was formed. It is an apt term. The dictionary defines "resonance" as a reinforcement or prolongation of a certain vibration through exposure to a similar vibration from another source. And because Sheldrake has examined other experiments that demonstrate this same principle of morphic resonance while trying to describe the accelerating formation of mineral crystals, he concludes that morphic resonance is a property of perhaps all physical and biological systems from the hydrogen atom right on through the human nervous system. Anywhere we encounter

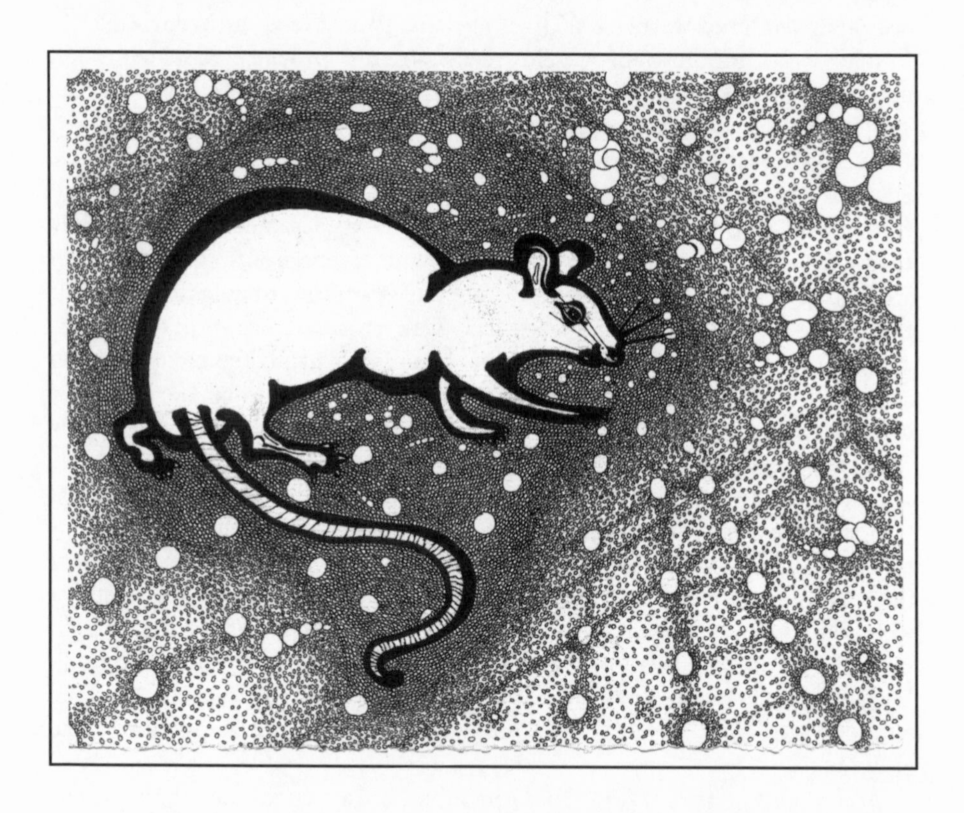

a nonrandom form, there we also find a morphogenetic field. Where we discover many such forms, there, also, exists the inevitability of morphic resonance.

One wonders if there are any limits to the scope of Sheldrake's formative causation. For example, since Gaia is best perceived as an organism, morphic resonance may contribute to the process of self-regulation. Consequently, all the individual "parts" of Gaia—you and me, the algae and the alligators, the methane-producing termites and the oxygen-producing rain forest—interpenetrate like the genes that work together to regulate any single organism. Gaia is our survival machine.

But why stop at Gaia? The universe itself is a nonrandom form that operates by mathematical principles. We may speculate (wildly or otherwise) that the evolution of the universe is a regulated process guided by some grand-scale morphic resonance. The big bang, that so-called first cause, may be depicted as the introduction of a morphogenetic field. Sheldrake is well aware that formative causation links biology and physics with concepts usually associated with religion and cosmic consciousness. After all, many religions describe God as the vibration of the universe. The potential of morphic resonance certainly adds new meaning to the words from John 1:1:

> In the beginning was the Word,
> and the Word was with God,
> and the Word was God.

Words are sounds, and of course, sounds are vibrations.

On the level of human beings, there are several key mind functions that may utilize formative causation. Telepathy might best be understood as a self-aware focusing of the "beam" of the usually subliminal morphogenetic field. Likewise, Jung's concept of dreaming and the collective unconscious may one day be described as a morphic resonance that unifies the species through subliminal archetyping and simultaneously regulates the evolution of the human mind in regard to Gaia. The aborigines must understand something that we do not. Why else would they name their totem relationship with nature, one that employs a well-tuned telepathic sense, the Dreamtime? Or for one more example: Suddenly the power of group prayer acquires new potency, whether a person worships God or is a dyed-in-the-wool atheist.

Earlier, I asked if you, the reader, were ever cognizant of a unifying field that somehow joins us all to each other as well as to everything else. How, in fact, do we develop and expand upon this sometimes fleeting awareness of "cosmic" consciousness? First, let us make a serious effort to demystify the experience, to stop representing it as a rarified and even secret ability. Second, and conversely, we must strip away its cultural patina of hocus-pocus, and stop treating it as the naive delusion of the superstitious. Third, we must stop sensationalizing the experience, as if it were the talent only of certain gifted individuals, some of whom use it as a mechanism for professional entertainment.

Instead, let us fully recognize that these inter-dimensional leaps of unifying consciousness are at once natural—of nature—and also, a potential link between the human individual and Gaia. In a very real sense, self-awareness of morphic resonance provides one possible route toward the healing of the earth. And it is no accident that we tend to label our own experiences "stopping the mind," or "stepping outside the mind." The terms are accurate; they serve to distinguish an experience that is uniquely distinct from the workings of day-to-day consciousness—the consciousness that remains so tightly bound up with the stuff of human culture.

At this particular juncture in our species' evolution, we humans are the only animals whose own creation—that which we cumulatively call "culture"—has fundamentally overshadowed the behavior of the biological being. Furthermore, and because of the immense destructive power inherent in some of our cultural products, the human-acculturated mind firmly charts the future course of both evolution and of Gaia herself. Edward T. Hall calls this sinister phenomenon *evolution through extension.* He writes:

> Both quantitatively and qualitatively, the gap is too great between man and the other species to say much more than, when the process begins, evolution accelerates drastically; therefore, because man is unique in this regard, he can learn only from himself.

The obvious way for us to temper and control the excesses of our acculturated mind is to develop techniques that allow us to "step outside our minds." Through meditation, rigorous exercise, conscious dreaming—in fact, through any number of disciplines—we learn to tap into that other mind. Here is the realm of morphic resonance, of unifying consciousness, of one-world community. Of natural wisdom.

In this light, Sheldrake's greatest contribution may be his thorough insistence that the scientific establishment become cognizant of a phenomenon that the mystics and primitives have known about all along. He has couched his hypothesis in the traditional and very precise language of the professional journals, thus lending to the concept a true scientific credibility in an age that believes in its science—and listens to its scientists. The experience of natural wisdom becomes available to more people than ever before.

The dolphins may prove valuable guides along the path of this conscious journey to realign ourselves with natural wisdom (which I now use interchangeably with both morphic resonance and unifying consciousness). Hopefully, Edward Hall was wrong after all; hopefully, humanity *can* still learn something from another species. And because of their unabashed friendliness toward humans, dolphins often seem like ambassadors from the otherwise secret wisdom world of nature. Of all the wild creatures, the large-brained dolphins seem the most willing to meet us halfway.

And although they usually appear totally inscrutable when viewed through human perception, in fact dolphins utilize natural wisdom from the very center of their consciousnesses. This may seem a brash statement, but I have felt its transmitted manifestation too often to believe differently. Their transmitters and receptors are a potent force indeed, most capable of piercing through even the skeptical human being's vast armory of acculturated defenses. It is indeed disarming (the perfect turn of phrase) to suddenly turn silly and awestruck, vitalized and yet secure, in the presence of dolphins. Most people don't have a clue as to what is going on inside themselves. Very often, I have heard it described as an exhilarating sense of unity with the dolphins. What we are feeling is the immediate. It is the immediacy of beings in resonance.

But I must stress that natural wisdom is by no means the exclusive province of the dolphins. It is the wisdom of the turkey and the buffalo, of the gray whale and the mitochondria. It is the pull of Gaia, a force that interpenetrates all of nature. I, for one, have felt it at its strongest in the presence of seventy-foot-long fin whales. The larger the creature, the larger its bioradio transmitter. Unfortunately, we humans have already decimated most of the larger whales. They are no longer readily accessible to a long-term exploration of the interface.

But dolphins are still common animals. More so, humans and dolphins often seem like opposite faces of the same large-brained coin. We tend to reflect each other. More of us humans sense natural wisdom more clearly and succinctly from the dolphins, than from any other order of animals. In such a light, it seems our own crucial responsibility to thoroughly commit time and energy to an exploration of the human/dolphin interface. Not to teach them English, not to test their intelligence, but rather to harmonize and resonate with them. And if we make that commitment during this generation, then just like McDougall's rats, our descendants and the dolphins' descendants will find it that much easier to plot the right course through the maze in future generations.

Author Ted Mooney, in his *Easy Travel to Other Planets,* provocatively speculates about some of the natural wisdom that humans and dolphins may share with each other. One dolphin thinks to himself:

> It was said that humans dream with their hands, only their hands, and so have cities rather than sagas, monuments rather than memories. He had reviewed in his mind every movement and gesture that the woman's hands had ever made within reach of his sonar, he recalled every dream their touch had caused his skin to dream—but still he did not understand what made her come and go, come and go from dry to moist to wet and back again, and what his name meant in her language and what his language meant in her

ear and where her offspring were and what hair was for and why she was afraid of the shortness of life.

I mean to offer Mooney's scenario as nothing more than an example of the kind of territory that we may soon be surveying on a day-to-day basis. What does Mooney mean when he describes the dolphins dreaming with their skin? There must be something to it. I once heard a Chumash Indian elder deliver a similar verdict about the connection between dolphin dreaming and skin.

Summer expeditions with the orcas at Buddy's Cove verify that a long-term interspecies community is not only a worthy venture but also an entirely realistic proposition. But Buddy's Cove remains beleaguered by too many whale researchers all working through their many varied agendas. For one group of researchers to initiate a deep trusting relationship between species would be a disservice to the orcas. There is simply too much potential for other humans to abuse the trust. Furthermore, and most critically, the ocean temperature around Buddy's Cove is much too cold to permit humans to swim with the orcas on a normalized basis. Also, the orcas seem to frequent the cove for only three months a year. As such, the human/orca community survives as a summertime event that flourishes at night and is preponderantly acoustic in nature. The surface of the water will forever exist as an insurmountable barrier that separates skin from skin. And skin to skin is one place we must go.

In 1982 I was awarded a grant by the Human/Dolphin Foundation to attempt to initiate a relationship with the free-swimming dolphins that were known to live off the coast of Careyes, a sunny tourist resort located a hundred miles south of Puerto Vallarta along the Pacific coast of Mexico. Specifically, my job involved attracting the resident dolphins into a small cove on a regular, predictable basis. This was planned as stage one of a much larger project that would eventually include a health spa and John Lilly's dolphin computer lab. All of us human participants agreed in advance that music and in-water play were to be the only enticements. At no time in the unfolding project, either now or in the future, would the dolphins be held captive. Once these ethical constraints were established, we budgeted ourselves six months just to see if it was even possible to develop an ongoing and daily relationship with dolphins, in a place where none had existed before.

My wife Katy and I arrived at Careyes in late November 1982. By the second week in December, our small motorboat was ready for service. And so, every morning at 9 A.M. we ventured forth into the huge bay, searching for dolphins.

We found a veritable neighborhood of leaping sailfish, giant manta rays, yellow and black striped sea snakes, sleek dolphin fish, translucent and improbably shaped ribbons of jellyfish, and migrating humpback whales, who filled our acoustic monitors with some of the loveliest songs in all of nature. Our sponsors were hoping that we would also find tursiops, bottlenose dolphins. These were the

species that John Lilly had come to know so well after many years of his own interspecies research. Tursiops are known to venture close to shore. They are the species best known to mingle with humans. The dolphins who venture to shore at Monkey Mia are tursiops.

Unfortunately, we saw tursiops on only one day. They were swimming quite close to the resort's main beach, seemingly scoping out the people who swam into their midst to play with them. But after a few turns around a small island near the beach, the dolphins exited after about a half hour. They were never seen again during the entire period of the project. However, on several occasions, we saw bottlenose dolphins cavorting close to shore at two large bays located about an hour south of Careyes. Both of these southern bays had very calm water, and a predictable surf. We concluded that the shoreline at Careyes, with its convoluted volcanic cliffs and foamy surf, was simply too rugged to accommodate a resident population of inshore tursiops.

Likewise, we also ran into a pod of very fast-swimming pseudorcas—once heading north—and two weeks later heading south again. They were big, powerful animals, as vocally gregarious as the orcas of Buddy's Cove, and just as imaginative in their on-pitch, improvisatory replies to my guitar melodies. We hoped that they would linger within the bay for at least a few days. But they were soon gone, never to be seen again.

Farther from shore, at least a mile out, we often met up with pods of spinner dolphins. These are the same species that I had met with, years before, off the big island of Hawaii. Spinners are one of the two main species that die by the hundreds of thousands in the nets of the American tuna industry. Tuna fishing is a straightforward maneuver. For some reason, where there are large herds of dolphins swimming on the surface of the sea, there are inevitably tuna just below and out of sight. The fishermen surround the dolphin pods with a huge purse seine net. When the purse is drawn tight, the tuna swimming below the dolphins are ensnared. When the fishermen pull in the net, the dolphins drown. This needless slaughter is a result of fairly new and deemed (by the industry) *improved* technology. Before 1960, the tuna industry used fishing lines to catch tuna one at a time. No dolphins were caught. Some estimates place the incidental dolphin kill since 1960 well into the *millions.* This needless slaughter is why so many environmental groups plead with their members to stop eating canned white-meat tuna, which is caught with nets that also snare dolphins. Light meat tuna is caught with hooks and lines.

Since spinner dolphins are known to congregate in fairly large herds, we often wondered if the small pods that we saw so often, usually six to fifteen animals, might not be the remnants of formerly large tribes now seeking a safe haven closer to the coast. We also wondered if their hesitancy to venture close to our boat had anything to do with the genocide of their species that still occurred at the hands of other humans out on the open sea.

On only one single occasion did we achieve what might be called an interaction with the elusive spinners. It was December 29. First, Katy and I traveled alongside two humpback whales, who were slowly plying their way south. Just as we began to serenade them with music played on a bamboo flute, one of the whales jumped completely out of the water just fifty feet from the boat. They continued south, and finally were gone. A half hour later we watched a sailfish leap diagonally into the air at least twenty times in a row before it finally settled down. Then a pod of spinners appeared just off the point that defined the southern extremity of Careyes Bay. They swam around the boat several times. I entered the water. They, in turn, repeatedly swam just within underwater visual distance—about fifty feet—and for about a half hour. Then Katy also entered the water. They circled us, always skittish, always keeping the same distance from us. But, curious nonetheless. We left the area only after it began to get dark. The spinners traveled alongside the boat for about a mile as we traversed the bay back to our mooring in front of the resort. Then they were gone.

But like the tursiops, the pseudorcas, and the humpback whales, the spinner dolphins were not resident to Careyes Bay. Sometimes we would go for weeks without seeing them. Once, we were very sure that we had met up with a new group, because one member sported a bright white "freckle" on the top of his dorsal fin.

The dolphins that we *did* see regularly, usually about a mile offshore, were called spotted dolphins (*Stenella attenuata*). The spotteds are the other victim of the American tuna industry. Over time, we came to realize that we were very often meeting up with a group of six spotted dolphins. And despite the fact that we were never able to make positive visual identification, their behavior toward us changed so dramatically for the better over several months' time, that we became convinced that it was the same group. Once in a while there were eight of them—the usual six, plus an evidently independent female and her tiny but very speedy baby. What a sight it is to watch a two-foot-long dolphin leap completely out of the water. We had found our resident pod.

The issue of residency was, of course, critical to the success of the long-term goals of the project. The dolphins who interacted with us today had to be the same dolphins who interacted yesterday. Otherwise, it would become impossible to achieve any sort of continuity to the relationship.

Our first six weeks out on the water were devoted to establishing a basic working method for approaching the skittish spotted dolphins. There were simply no precedents to guide us. Nothing quite like this human/dolphin community had ever been tried before. In retrospect, it seemed as if we ended up trying much too hard to accomplish the task at hand. Every morning we'd load up the boat with sound gear, masks, snorkel and fins, sunburn cream, and a few pieces of tropical fruit, and proceed to go searching for dolphins. The boat was very fast and so it covered an awful lot of territory very quickly. Usually, within the first hour, we would spy the mottled dorsal fins of the spotted dolphins. They were usually swimming along, seemingly minding their own business, neither interested nor disinterested in this motorboat so suddenly in their midst. We stopped the boat, plugged in all the wires connecting hydrophone to headphones, and musical instruments to underwater speaker, and began to play whatever music entered our heads at the time. Occasionally the dolphins would stop whatever it was they were already doing, and venture close to the boat. Then the echolocating began in earnest—very different from the strident pulses of an orca—more like the sonic marriage of a creaky door and a rattlesnake. Now I began to imitate dolphin-style whistles on the soprano guitar by carefully stroking the high strings with a bottleneck while carefully modulating the attack and decay of each note. Katy put on mask, snorkel, and fins, and jumped into the water. Now the dolphins were within a hundred feet of the boat. This, in turn, elicited more echolocation. Some of the bolder dolphins began to dart back and forth within ten feet from Katy, although she could not always keep track of their swift movements while in the water.

Although it was always exciting, we also knew that the dolphins were extremely wary of these strange humans who, every day, ventured out to interact with them. Usually after five minutes or so, the dolphins would regroup, and go on their merry way. Katy would quickly climb back into the boat while I, in turn, would uplift the hydrophone and speaker over the gunwales. We'd start the engine, and zoom after the dolphins. Then, we'd cruise alongside the pod for a few minutes while trying to decide if we should chance losing them by shutting off the motor. Finally we'd stop the boat. Sometimes the dolphins would stop as well, and so the entire process would be repeated again. Rarely, these stops and starts might go on for three or four passes before the dolphins finally made up their minds to disappear. But just as often, the dolphins would dive as soon as we started the engine. They were easily able to evade any further pursuit by exiting underwater. In general, we were never able to tell if all this activity was having any positive effect at all. And if we had learned anything about how to get closer to dolphins, it was that they ventured closer only when the human in the water was entirely submerged.

After six weeks of chasing dolphins, we knew that it was time to drastically overhaul our basic working method. First, we agreed to cease all aggressive motorized behavior. It was too blatantly intrusive, and thus counterproductive to the ultimate goal of community. Reading through our notes, we also came to realize that no single in-water event, no matter how exciting it seemed at the time, had ever lasted more than ten minutes. We agreed that it was probably better to sit in the boat until that time when the dolphins themselves gave us some sign that they desired such close contact. Yet also, we knew for sure that the dolphins were beginning to respond to some of our music. But what kind of music was best? We weren't sure. We agreed to fully concentrate our attentions to developing the music. In conclusion, we opted to allow the dolphins themselves to set the terms for this relationship. Naturally, it was at that point in the proceedings that we finally felt the first glimmerings of a true interaction. Things began to happen.

First, in early January, we chose a specific meeting place, a *svaha*. It was a spot about a mile offshore, directly out from the cove where we eventually hoped to attract the dolphins. By this time, we had observed enough of their behavior to feel convinced that these spotteds would never venture close to shore by their own volition. It was contrary to their basic nature. If we were ever to lead them to the beach, it would come only through a long, patient process of instilling trust. So we chose the svaha. We played music from that spot whether the dolphins arrived or not. And if they showed up for a while and then left, we never chased after them. Instead, we sat quietly or went fishing for a while, and then returned to play once again. Once we had chosen the svaha, and so headed straight for it every time we went out on the water, the dolphins very quickly realized as well that it was to be a special meeting place. From then on, until the conclusion of the project two months later, interactions occurred over 80 percent of the time. During the last two weeks of the project, we met them there almost 100 percent of the time. More than once, the dolphins were waiting at the svaha when we arrived at 9 A.M.

Second, the dolphins invented a game. This consisted of swimming toward the boat and then quickly away again—back and forth like a yo-yo—each successive pass bringing them progressively closer to us. Shortly after our pleasant realization that this behavior was, indeed, a game, by every measure of the term, we started to lend musical accompaniment to the maneuver. It was simple: I played music until the dolphins reached an arbitrary line-of-sight distance from the boat and then I stopped playing. Soon enough, the dolphins caught on that this silence meant that it was time to turn around again. When they did, I'd start playing. But I do not mean to imply here that this communicated game ever became a set trick to be pulled out of my musical hat at will. Nothing in our relationship, except time and place, ever became routine. I call this conscious behavior a game because it involved innovative variations on a communicated theme. It was always dynamic, and yes, very much fun.

On one improbable occasion, the entire back-and-forth game occurred without Katy's or my ever laying eyes on the dolphins. They swam back and forth entirely underwater. Yet it remained easy for us to monitor their relative distance from the boat by noting the increased and then decreased volume of the whistling. But by this time in the relationship, the thought passed through my mind that the dolphins were having a bit of a practical joke on us. That is, they were not swimming back and forth at all, but rather, whistling louder and then softer.

Third, I phased out my poor attempt at mimicking their whistles. I reasoned that most of their vocal content was being spoken in a frequency range far above the range of either my ears or the electric guitar. All that I was actually copying, even at my most inspired moments, was the low end of a wide-band whistle, which in its entirety was unimaginably rich. For all I knew, those whistles may have been the dolphin equivalent of poetry. It was as if I were to attempt to reproduce a Beethoven symphony after one listening by mimicking a single violin part on the electric guitar.

So instead, I began to experiment with just about every sound that could be plucked, struck, stroked, rubbed, tapped, or bonked out of that little Peavey guitar. Sometimes I played songs, favoring my orca-reggae boogie, and a chord melody of that old standard "Misty." It was slow going. No particular tone or tune seemed to matter much as long as I did not play too loud or too often.

Finally, a breakthrough occurred. It was early February; the weather had recently turned colder—sometimes as low as 65 degrees when we ventured forth in the morning. As we arrived at the svaha one morning and lowered the acoustic gear overboard, I listened to a new and very distinctive sound emerge from the usual hiss of the ocean. It was percussive, and quite random—sounding like a room full of fast typists, if that can be imagined in two hundred feet of water. We saw no dolphins that day, nor did we hear any whistling. It was easy to mimic the sound by dampening all the guitar strings with my left hand while strumming the strings with my right. Whether I played or stopped, the random clicking remained exactly the same.

That evening I sought out a local fisherman, who knew more about the underwater environment than anyone else in Careyes. He chuckled and told me that we had been listening to a species of bottom fish who became vocal during that time of year. He called them "grunts," although he added that he was not sure they were "officially" members of the grunt family.

When we returned to the svaha the next morning, the ocean was silent except for the constant *hisssssss.* We sat quietly in the boat for fifteen minutes, listening intently, either for dolphins or for the new sound that we had quickly nicknamed the "fish rhythm." Finally, I cautiously began to play my rough approximation of the new sound. Not surprisingly, within just a minute, the ocean became alive with the sound of "typing" bottom fish. Together, we clicked for the better part of the next hour. Then, from very far away, we began to hear the distinctive whistling of dolphins. The spotteds were coming. They arrived just a few minutes later. All the time I continued to mimic the muffled fish rhythm, although I found it increasingly difficult to keep it random. Instead, it began to emerge as a precise *one*-two-*one*-two. I played for twenty seconds, and then stayed silent for twenty seconds. Over and over again.

Suddenly, a new and distinctive sound issued from out of the ocean through my headphones. Like the guitar, and like the bottom fish, it was a variation on a basic percussive clicking. But this new voice was much more sharply defined than the fish rhythm. It sounded less like a mad typist and more like a human percussionist striking together a pair of, well . . . dolphin sticks! My busy mind was quickly filled with visions of Jackson Jacobs calling the dolphins to shore by striking together his dolphin sticks. It all became very clear. Everybody—dolphins, humans, bottom fish, shamans, guitar players—everybody was playing music with everybody else. The vision of nature as a band at a costume ball. Play on!

Although I had lost my own dolphin sticks just two months earlier, still I found it quite easy to replicate the sound by striking the damped guitar string with the edge rather than the flat of the plectrum. Or the simple matter of cutting the treble, joined to the very difficult matter of picking randomly, approximated a basic fish rhythm. Or, for the third variation, adding a bit more treble, adding a semblance of nonrandom rhythm, and utilizing the flat of the pick rather than the edge replicated a fair approximation of a dolphin mimicking a fish rhythm or a pair of dolphin sticks.

We were soon quite convinced that a dolphin had, indeed, produced the new sound. This fact was made plain when we next heard several clicks in a row. The sound was very loud and clear; it had to have been vocalized right next to the boat. I peered over the gunwales. And at just that moment, one of the spotteds surfaced and blew five feet from the stem. He looked right at me with that inscrutable dolphin grin. We had found our clicker.

The sound itself was a dolphin vocalization never before heard in two months of close observation. That it was invented right on the heels of both the fish rhythm and the guitar clicking suggested one of three things. First, that we were projecting our interpretations too much. It was probably coincidental, and "meant" absolutely nothing. Second, that the dolphins naturally vocalized this sound every February in imitation of the bottom fish. Or third, that the dolphin was responding entirely to the guitar mimicking the bottom fish.

Over time this special sound developed into our main form of dolphin/guitar signaling. Furthermore, and most significantly, it was one of the very few sounds that both guitar player and dolphin could produce and hear as well. It was not coincidence. It may have been what John Lilly so mischievously calls "coincidence control." Through the random genius of the bottom fish, who vocalize every February, the humans and the dolphins had stumbled upon the fragile beginnings of a mutually acceptable musical language. The sound itself was unique to our developing relationship.

Less than an hour after we had first heard the dolphin's rendition of a fish rhythm, it became clear that there were three dolphins playing around the boat, and that each one possessed an individually identifiable voice. One voice in particular stood out from the other two. It was slightly deeper and more resonant, and it seemed to be leading the other two. Several times, the low-voiced dolphin initiated a sequence of fish rhythms, which the other two repeated in nearly chorus-like precision. When I first heard this "song," I decided to remain silent—to listen and learn for a future encounter.

The next day, after I had played guitar fish rhythms for about a half hour, five dolphins swam into sight for what was to prove quite a lengthy session. The fish were clicking; I was clicking back: *one*-two-*one*-two. Suddenly the low-voiced dolphin began an incredibly precise meter: *one*-two-*three*-four—onetwothree*four,* repeated over and over again. Furthermore, the rhythm itself seemed to be based in the key of D-major. Inexplicably, and without a thought as to how I might best answer the dolphin, I began to play "Misty," also in the key of D.

It continued this way for about a half an hour: the fish droning away like the background to an Indian raga; the lead dolphin keeping a precise *one*-two-*three*-four—onetwothree*four;* one or another of the other dolphins repeating that rhythm in chorus; other dolphins whistling; and a guitarist breezing through the strains of "Misty." If nothing else, the fish, the dolphins, and I had achieved the beginnings of a band that might be able to make a living playing weddings and bar mitzvahs.

The next morning brought another noticeable change to the dolphins' fish rhythm. Again, with a low-voiced dolphin in the lead, ostensibly the same animal, he or she struck out a new variation on yesterday's rhythm:

*One*-two-*three*-four——onetwothreefour*five*

Once again I began playing "Misty," but this time trying to mold the melody around this new and tricky rhythm. The result may have no longer sounded like mellow cocktail hour music, but it was still as recognizable as ever. It sounded like something that might make Dave Brubeck smile and snap his fingers.

Now the dolphins began to play their back-and-forth game—rushing off at top speed, all five breaking across the surface of the water in unison, away from the boat, until I stopped playing. Then they quickly turned about, and zoomed back to the boat. Then back and forth a third time. On this third pass to the boat, Katy dropped into the water, and swam with the dolphins for the first time in over a month. And for the very first time in the entire duration of the project, the dolphins expressed no fear of her presence. To the contrary, they swam right up to her, echolocated, copied some of her in-water movements, and then swam off for still another back-and-forth pass. In toto, that morning, they made three return passes back to Katy while she swam in the water. It was the first time a dolphin had actually returned to a swimmer of its own volition. It was another breakthrough.

On that day, Katy and I finally became convinced that, given enough time, we would eventually succeed in establishing a human/dolphin community where music and play were the reasons for the bond. However, we had no clear idea how long it would take, if it would ever move into shore, or if it would develop in any predictable manner. It was up to the dolphins to lead us for a while.

Over the next month of the project, this process of swapping rhythms and creating melodies over varying dolphin rhythms evolved into an increasingly healthy relationship. More human swimmers were added to the back-and-forth game. One day, in particular, stands out from the rest. The dolphins were waiting for us as we arrived at the svaha. Then, the low-voiced dolphin initiated one of his standard four-beat fish rhythms. He called but one single time. Then, the entire pod quickly swam off, not to be seen again that day. It seemed a signal that yes, indeed, they had kept the appointment, but needed to be somewhere else that morning. The following chart outlines some of the story of our relationship:

A. *Days out on the water.*

B. *Dolphin sightings:* This refers to visual contact. Any group of dolphins is counted as one sighting, regardless of duration. If they leave our sight, and then later return, it is counted as a second sighting.

C. *Dolphins heard:* Refers to auditory contact through a hydrophone. If reception takes place in a particular location, no matter the duration, it is counted as one contact.

C. *Rapport established:* Refers to the dolphins responding to our presence, whether visually or acoustically or both. A true interaction occurs. Examples: Dolphins change their course to meet the boat, interact with humans in the water, repeat musical or numerical patterns, improvise a dialogue through clear call and response.

| | December | January | February |
|---|---|---|---|
| A | 16 | 18 | 17 |
| B | 10 | 12 | 10 |
| C | 1 | 13 | 16 |
| D | 6 | 10 | 16 |

In late February, a Mexican film crew arrived at Careyes to document this germinating human/dolphin community. Interestingly enough, that particular morning, eight people swam with the eight spotteds for more than two hours. It turned out to be one of the very best interactive occasions of the entire project. Even then, it was quite evident that the dolphins swam closest to Katy, who was by now well known to them as a responsible playmate. The completed TV documentary was later aired to 130 million people throughout Latin America.

However true the relationship seemed to be ringing, still there was serious trouble brewing for the future of the Careyes Project. This was due to that all-too-common problem of differing human expectations. After nearly four months of dedicated work, the needs and the goals of the various human participants had changed. Old issues metamorphosed into insurmountable problems; new issues led to major differences of opinion. Much to our chagrin, and despite our growing attachment to our friends the spotteds, Katy and I left the project by the end of March.

First, despite our successes with the Careyes dolphins, spotteds are characteristically shy beings, not possessed of the bold amicability toward humans that so distinguishes their larger cousins the bottlenose, the pseudorca, and the orca, to name the better-known species.

Second was the lingering issue that the spotteds never ventured close enough to shore to allow for the long-term project to evolve into its next phase. Several sponsors had initiated the Careyes Project because they wanted a dolphin language facility built on the beach. It was clear to everyone involved that this would never come to pass in the allotted six months. Of course, Katy and I openly wondered if even this problem might not solve itself in, say, a full year. But realistically, given the expense and the risk involved, it was mutually decided that this was reason enough to look elsewhere.

Third was the issue that the beaches themselves were not truly conducive to near-shore human/dolphin interaction. The surf was simply too rough to permit easy and continuous access by children. A school where human children might mingle with dolphins as part of their curriculum was one cherished facet of the larger dream of community. Play is pleasurable; it is its own reward. The dolphins are masters of play. Such a school would eschew regimentation to fixed schedules, and instead would be guided by the time when the dolphins chose to interact with the children. A school based on dolphin time. And how might a child then look upon the world, behave politically, make decisions, order priorities, organize his or her life, and last but not least, how might this child actually think as a member of a human/dolphin community?

Fourth was the critical issue of community itself. None of us would ever be able to establish a human/dolphin community unless we humans involved were ourselves a community. It is one thing to design a research project, and another thing entirely to design a community. Once we began to achieve our first clear sign from the dolphins that the relationship was developing, we knew that it would be different from the original game plan. Yet the human social situation of keeping to that original game plan became so politically complex and ambiguous that all the human principles began to read into the reality whatever our own goals—about dolphins, research, community—suggested.

Some people wanted to *study* dolphins. They were as concerned with the way we kept our notes as they were about the relationship. Other people simply wanted to *be* with dolphins. Even this meant different things to different people. Some sought an esoteric *peak experience* that might be attained simply by spending time in the presence of dolphins. Other people were aligned with the resort hotel of Careyes. They envisioned the developing bond as a kind of new-age *tourist attraction.* The spa designers looked upon the dolphins as *healers.* Personally, after nearly four months in daily relationship with dolphins, Katy and I felt that this idea demanded much consideration. We did not yet know how to develop this aspect. Unfortunately, the spa designers sometimes talked about the dolphins as something to partake of between a 10 A.M. Jacuzzi and a 12:30 dose of megavitamins.

But I do not mean to sound cynical here. Every single idea had its merit. Ultimately, we were all beginners. In fact, during the entire four months of the project, it became a very high priority for us to include interested people in the daily process as observers, participants, strategists, and swimmers. Some sat in the bottom of the boat and watched in awe as the dolphins came in, or even if a manta ray jumped. Others became seasick just as the dolphins arrived, forcing the boat to return to shore. Other people insisted that we abandon our electronics in favor of bamboo flutes and telepathy. Still others, including musician Terry Riley, played upon the tiny Yamaha keyboard and initiated their own human/dolphin interactions.

Because we did believe in community, we encouraged the input of all people interested enough in the concept to make the long journey down to the project site. Every single person offered original insights and activities. Two specific events stand out in my mind—probably the best and the worst of the lot.

The first occurred when we took Fred Stern, an author of computer books, and his wife, Betsy, out for the morning. Fred, all six feet five of him, lay face up on the floor of the boat with a hat covering his eyes. I had been playing the guitar for about a half an hour, without any results. When I finally stopped, Fred removed the hat and stated matter-of-factly: "The dolphins just told me that three of them will be arriving from the south in ten minutes." Then he paused a moment before adding, "Gee, I hope that what I inferred as ten minutes is what they actually meant." He sat up abruptly, drew the visor of his baseball cap down to protect his eyes from the intense sun, and continued: "The dolphins hit me with a 'block of time.' It wasn't exactly 'ten,' nor was it exactly in 'minutes.' I wonder if I interpreted their time correctly." He received his answer fifteen minutes later, when three dolphins suddenly appeared from the south.

Then there was the sulky, early January afternoon during an especially huge run of bonito. Both spinners and spotteds were feeding on them, no more than six hundred yards offshore, cavorting all over the surface of the ocean. A man had insisted that we take him, his girlfriend, and his son out to see the dolphins. We consented. But when the dolphins seemed to pay no attention to us, despite the fact that we were very near them, the man bodily picked up his seven-year-old son and threw him into the water. We were stunned. Why?!? "So that the boy can get a dolphin experience that he'll never forget as long as he lives" was the quick answer. That was indubitably true, because the little boy could barely swim despite his life jacket. He practically went into shock until we fished him out of the water. Only then did we realize that the man had had one too many margaritas. He said: "Don't you know that dolphins are well known for rescuing humans?"

On the one hand, we encouraged people's input into the project, yet on the other hand we were soon inundated with too many visitors for us to keep to our daily regimen. Because of all the varied and sometimes opposing ideas of what the project was supposed to produce, we suffered from too much public exposure, too early. Inevitably, we lost control of the entire mood of the relationship.

So once the musical and in-water play methods were found to be tried-and-true, we decided to leave Careyes. And if we felt dismay over leaving our friends the spotted dolphins, we were, if anything, ever more resolved to continue to seek the means to manifest the human/dolphin community. By now, we had a much better idea about how to proceed.

We need to focus much more attention on the issue of human community—of working with a group of people who share the common vision, and who also possess the necessary skills and personalities to make it happen. All together, we need to search out a location, in proximity to certain species of cetaceans, and then start the same process of interaction all over again. But, of course, we must make every effort to keep ourselves from getting stuck in a predetermined game plan. As such, we had better acknowledge, right from the start, that the dolphins are active and equal collaborators in the process. That means learning to live on dolphin time. The project will take its own shape.

After the relationship has established its basic tenor, there are many possible avenues for formal exploration of the interface. We need to film the community as a document of how one group of people has chosen to live in synchrony with natural rhythms. We might, in collaboration with Rupert Sheldrake, design any number of research studies that explore the resonant interface. This could include a scientific study of telepathy as a vehicle for interspecies communication. It might also include the relationship's potential as a medium of healing—both for the individual and for the planet. Why is it that people become so animated, excited, and yes, happy, in the presence of dolphins? I would like to set up a program that would bring talented musicians into the community. Given the state of human society today, a period of time spent with free-swimming cetaceans can provide a transformational experience for human beings. This is the concept we call "the dolphin as benefactor."

The concept of dolphin as benefactor is key to the enunciation of any future proposal. People from around the world who are involved in positions of leadership within their own sphere—whether it be politics, science, the arts, religion, business—are offered a retreat situation in daily relationship with free-swimming dolphins. Such an experience has already proved itself positively transformational for those who permit it to happen. It serves as an intellectual provocation, a source of joy, and especially, a profound connection with the natural world. Everything is connected to everything else. These are valuable lessons for our leaders to possess as we plunge and stumble into the twenty-first century. In my flights of interspecies fancy, I envision an opportunity for Koko the gorilla to meet the dolphins.

We transcend the power of dolphins as flesh and blood animals, and so, engage them as metaphor: a bridge capable of returning us to the ways and means of Gaia. And if they do possess any great or special power, then it is best explained as their unerring ability to demonstrate for us our own natural wisdom.

I am reminded of what is so often written about the great teachers of Asia. This teacher can teach you how to levitate. That one over there can show you how to tell the future. A third has lived for ten thousand years. But the greatest teacher demonstrates for you your own true nature. Animals do not tell us to be happy, never demand that we dwell in nature. Rather, often in their presence, we are happy, we do dwell in nature. On one level, their great gift falls away as nothing more than our permitting them to act as benefactors. But there is nothing matter-of-fact about this gift. The gift given by animals is precious: a guide back to balance. On that level the gift is the basis of a profound mystery.

This juxtaposition of natural wisdom, animal teachers, and mystery is where we began this book. It is the relationship known as totem. It is animal dreaming.

# Epilogue

ALTHOUGH THIS BOOK was first published in the mid-1980s (as *Dolphin Dreamtime*), I still get mail from readers asking me where they can go to spend personal time with dolphins. But whether they are seeking dolphin-assisted therapy, petting pools, swim programs, psychic workshops, underwater birthing, or dolphin language labs, I end up telling them that exploiting cetaceans for human ends, no matter how critical it may seem to their personal development, is not what I had in mind when I wrote the book.

While I will always honor the dolphins for leading me to the edge of the Dreamtime experience—a profound revelation about the human place within nature—my communication research parted ways with the smaller dolphins long ago. The bottlenose, the spinner, and the spotted dolphin, species of the smiling mouth and the bulbous forehead, vocalize primarily in frequencies far too high for a human musician to hear. Finally realizing that I would never be able to hear their calls without the aid of computers, I decided there was no point in trying to improvise music with them. So I began playing guitar instead with orcas, who vocalize in about the same frequency range as a guitar and whose calls possess much more innate musicality than those of any beaked dolphin. I spent twelve years researching communication with orcas.

My original emphasis in writing this book was not specifically promoting dolphins, but rather, confronting the Dreamtime, that indigenous Australian concept of allowing nature's creatures and habitat to serve as guides for reconnecting with a sacred sense of time. At the moment of writing the book, dolphins were serving as my guide. I had just finished directing a successful interspecies communication project with John Lilly off the Pacific coast of Mexico during which we enticed spinner dolphins to swim every day to a set place at a set time. I had also made personal contact with Australian Aboriginal people who responded to my dolphin communication work by demonstrating for me their own totemic relationship with bottlenose dolphins. Lastly, I had recently directed a two-year project for Greenpeace at Iki Island, Japan, employing activist strategies in an attempt to stop the killing of thousands of pseudorcas, grampus, and bottlenose dolphins. Local fishermen had over-fished local waters. With their traditional economy headed for a crash, they had started slaughtering dolphins in a futile attempt to develop a new "fishery."

With these experiences fresh in my mind, I wrote *The Man Who Talks to Whales* to bear witness to the sublime effect dolphins had on my psyche, my environmental activism, and especially my deepest feelings about living within nature. Are dolphins animals? Certainly, but they were also a non-human people. Sensing the truth of that conclusion, it was an easy next step to recognize that all animals are non-human people.

Despite the effect that dolphins had on everyone I knew who came into contact with them, I always conceived of the relationship as a metaphor for the more radical premise that animals are people: sentient beings. *The Man Who Talks to Whales* served as a guide for readers to find their own method, their own metaphors, their own totem, their own experience, to guide perceptions, to help them re-perceive wild nature as culture, as community. When we are finally able to sense that linkage on a cellular level, we become skilled enough to gather up the courage and then the resources to teach the rest of lagging humanity just how much, at this moment in history, nature needs its human beings to comprehend this humility called Dreamtime.

Nothing in this book declares that a reader's personal salvation is at stake unless he or she signs up for a guided swim program in Florida or Hawaii. And certainly there is nothing that urges a reader to go book a room in a fancy resort that will, for a slight fee, allow your kid to pet dolphins who get tossed a fish in payment. The dolphins don't need human beings impacting their lives. If you love them well you will love them dutifully; and you will do nothing besides helping them to prosper. Ironically, you may aid them best by developing some other Dreamtime guide. Love the one you're with. What, you ask, could possibly be a more potent Dreamtime guide than the large-brained, super-energetic animal-celebrity dolphins? Look outside your window. You might see robins, ravens, maple trees. Rocks. Deer. Clouds. You get the idea. Allow me a story to explain.

## The Poppy Dreamtime

Late one summer I planted three Oriental poppy plants in three different locations within my sprawling garden. The first grew under the sparse shade of a young plum tree, in rocky soil amid plantings of sweet william and Canterbury bells. It seemed fairly happy in its well-drained location through the mild winter that followed. By mid-spring, however, it appeared to shrink as the nearby biennials suddenly shot up to four feet tall.

The second grew just fifty feet away in a bed raised above the flat mossy-covered bedrock that constitutes my front yard. Here, the conditions were quite different. This one spent the mild but wet winter in slow-to-drain compost. The moisture-loving poppy seemed to flourish.

The third one was planted well out of sight of the other two: down a hill, at the edge of a fir forest, in rusty-colored soil composed of a thousand years of partially decomposed conifer needles amended with wood ashes and peat. By the time of the first frost of late October, this poppy had easily doubled in size, then seemed to stop growing.

On May 1, plants one and three were about the same size while number two, growing in its composted puddle, was not only twice the size, but displayed double the number of flower buds as well. On May 20, precisely between 10 A.M. and noon, the first papery-textured flowers from plants one and two opened as if they had chosen one another as partners in a well-orchestrated and magnificently scheduled samba. Literally, they bloomed as mirror images of one another. Each flower was eight inches across!

I distinctly remember watching the two distant plants open their flowers in tandem, making me feel like a voyeur staring through Venetian blinds at a couple making love. I walked down the hill to stare soberly at the third of the triumvirate. It stood reaching toward the sun as a virtual triplet to the others, but an entire week away from joining them in the dance. What could have caused the two plants to show their first flowers of the year at the same exact moment?

We humans love to find explanations. We build our civilizations upon them. But how dare I posture; without the fruits of human knowledge I wouldn't even be gazing upon these exquisitely silken, apricot-colored cultivars from Asia. I noted that each of the three poppies was growing in a different place, so I felt safe to discount the soil, the water, and even the light as a motivating factor. Actually, the primary difference I was able to discern between poppies one and two on the one hand, and number three on the other, was the undeniable fact that the two samba dancers had been in view of one another every single day of their lives.

The poetry of this observation implanted itself firmly in my consciousness, undulating gracefully within the breeze of its plausibility until finally it blossomed forth with the stature of personal acceptance. What am I talking about? Do plants see one another? Do they dance? Are they creative? I offer no objective answer, although in this case the poppies did seem to be utilizing some kind of subtle perception of proximity as a means to blossom in perfect synchronicity. How do any of us come to dance with one another, until first we place ourselves in an appropriate position (sometimes like a flower), and then attain a state of mind (sometimes like a human being) to reflect the overwhelming sensations that spring from the wellspring of intimacy? Is that it? Poppies one and two were in love? Poppy three acted as an inadvertent wallflower?

Actually, trying to say what I mean makes me feel I have already lost my way. I attempted to get my bearings by returning to my original seat between the blooming poppies. I dropped down on the moss and gazed long and hard at the first one. Then I sat up, turned ninety degrees, and watched the other.

Both poppies lean into the sun, slowly surging from east to west in a slight breeze. The larger of the two looks as if it won't open any more flowers for at least a few more days, although the smaller plant could conceivably open three or four more by tomorrow. I look back and forth from one to the other. But wait, something is amiss. I am sitting directly between their lines of sight. Joining their dance, I rise to take two giant steps backward.

So here I sit upon the earth, a captive audience to the dance, and sanction a brass band of intellectual explanations to parade through my mind. Yet the longer I stay put, only one idea seems to make any sense to me. It speaks more to my role as audience than to the poppies' role as synchronized dancers. It is an ancient idea that delineates a linkage with nature, perhaps best expressed through a quote from a long-dead Lakota Indian chief named Luther Standing Bear:

> That is why the old Indian sits upon the Earth instead of propping himself up and away from its life-giving forces. For him to sit or lie upon the ground is to be able to think more clearly and feel more keenly. . . . Man's heart away from nature becomes hard; he knew that lack of respect for growing living things soon led to lack of respect for humans too.

The concept of the Dreamtime starts from the premise that animals and plants, habitat, weather, and even geology possess unique awareness, intuition, perception. As the poppies teach, the lesson includes love and wisdom as well. Accepting the Dreamtime means we permit poppies to dance with one other as well as with any human being willing to sit on the ground long enough to learn the proper way to perceive them. It's not such a risky business if you happen to be, for instance, a nineteenth-century Indian chief. However, no Lakota chief was ever accused of aiding or abetting a clear-cut through a rain forest or running an oil tanker aground in Prince William Sound. Any person who perceives nature as primarily a source of love and wisdom, and let's include health as well, is not likely to lend a hand in its destruction.

Luther Standing Bear tells us that nature exists as the recipient of our unconditional love. Take care, he murmurs from the grave. Each time we harm nature, we cut a little piece out of our own flesh. When we drive sea turtles, condors, and elephants to the brink of extinction, we brutalize ourselves in the process. The connections interpenetrate. Yes, that explains the Dreamtime. It is the time when connections interpenetrate, the suspension in time that Gary Snyder refers to as "everywhen."

Most human beings subscribe instead to the standard past, present, and future. In its own way, this triad creates the contemporary schism that discriminates between human social relations and natural relations as the only practical way to progress in this crazy world of ours. I turn again to face old Luther sitting contentedly inside my head, and imagine him throwing some dirt up into the air before commenting that we are simply deluding ourselves. The future does not exist because, simply put, it has not yet come into existence. The past is certainly the source of all our knowledge, but it does not exist, either. The present does exist, although only as a boundary between past and future without any duration of its own. Modern life is thus based on a system of time that is a delusion. Because nature exists separate from this system, the schizophrenic relationship to nature that human beings happily accept as "modern life" is a trick of perception. How is it possible to ever be whole, while living separate from nature's time?

Luther and I agree that the *everywhen* perception of nature informs a spiritual and unifying linkage to the ecology of our surroundings. This spiritual ecology is an ancient path, a spiritual path, here referred to by its aboriginal name of the Dreamtime. I could have called it something else. It is, after all, a perception, not a name. The temporal basis of this perception offers an obvious handle for musicians who create sound in time.

But this book is not specifically about metaphysics or music any more than it is about dolphins. Rather, I am promoting new perceptions about how to reconnect with nature through anecdote, myth, and story. It suggests we start granting dolphins due process under our laws, give homage to buffalo, treat spiders and Oriental poppies as co-creators and teachers.

If this Dreamtime is starting to sound unmistakably positive, why is it considered so hopelessly unattainable (or just plain naïve) in terms of the current worldview that drives our culture? How has it come to pass that we perceive of spirituality and nature as mutually exclusive items?

Objectivity, as it is utilized by the earth sciences of this early twenty-first century, treats us as gods: as if we are somehow able to view nature from the outside looking in. This separation turns us invincible, justified to manipulate and extirpate animals and environments in the name of that just cause known as the pursuit of knowledge. To utilize just one example: objectivity permits the medical research community to name a rat a specimen, an object, a term that justifies the injection of carcinogenic pesticides into the rat's body until it expires. The human dressed in his white lab coat designs an ingenious computer model, takes precise notes, publishes his results in a worthy journal, and so adds his results to that ever expanding ream of information concerning the flesh of mammals wracked by tumors. How easy it is for the observer to throw his used-up specimen into an incinerator to make room for the next rat.

What do we learn from the rat's sacrifice? We gain knowledge of the precise amount of synthetic poison that it takes to kill a rat (for some reason it is not enough to know that this substance was never meant to exist inside the body of a living creature). We then extrapolate the commensurate dosage that will kill a human being. The substance wins a qualified approval from a regulatory agency, permitting it to be utilized within the ecosystem in certain strengths below that amount.

We end up spraying it on agricultural products, likewise poisoning the soil, the insects, the birds, and the rats, and maybe even a few unborn farm workers as well. The food is harvested, scrubbed and packaged, and eventually fed to us and our children. Eventually, every creature living on this planet ends up with residues of this chemical substance inside our bodies. But no researcher has yet accumulated enough data to pinpoint the complex connection that exists between all the varied residues reacting and interacting with one another inside all the wildly varying living organisms strewn in its path. No one disagrees that such a study would be impossible to do.

Meaning that our human sense of objectivity chooses criteria for safety as a function of equipment, technique, and the political clout of the testers. In that light, our objectivity is best perceived as a qualified objectivity, what might just as easily be called a qualified subjectivity. Meanwhile, every single day, more chemicals pass the tests and are "safely" sprayed onto our food and into the land. Likewise, these new tests demand more experiments upon more and more rats. We are finally confronted with an endless chain of little murders perpetrated in the name of increased productivity, which ends up causing big and slow murders all over the planet. In such a manner, laboratory experiments on animals primarily teach us how shortsighted and unethical our species has become in its pursuit of progress.

John Berger reminds us that, "whereas animals were once central to our perception of the world, now . . . they have been reduced to the margins of our lives." Extending this idea into the realm of music and nature, composer David Dunn writes:

> For many African people, the sounds of animals are not merely the calls of separate organisms. They are the voice of a spirit form resident in that individual but also present in all the members of that species. That spirit is like a persistent and collective intelligence that defies geographic separation. This concept not only is present in the beliefs of traditional religious practices, but also appears as an essential trait of domestic life.

The prevailing culture sees it a bit differently. Animals live on in the modern imagination via secondhand models of nature including TV documentaries, children's toys, and the kinetic sculpture we find at zoos. At those places where humans daily confront the natural environment, whether urban or rural, real live animals seem forever in retreat. And when we do not see them, do not relate to them, we soon forget the experience of living in a world in which they are part and parcel. Without that experience, that direct connection, we go about our lives simply resigned to the fact that the natural world is vanishing before our eyes. It's nature, not us, that's disappearing, right? And anyway, what can any one person do about it?

You know more than you experience. A fierce gale in some distant country momentarily breaks through the glut of information arriving through the TV set to your ears and eyes. You have never visited that country, may never go there. You listen to political pundits argue brilliantly about the value of your own country's aid program, and how it encourages the grand ideal of a global village. Yet any amount of self-reflection must make you painfully aware that your head has become a repository for too much information. Although you do well to acknowledge the koan about a butterfly flapping its wings in South America affecting weather patterns in your hometown, you must also consider the way a secondhand perception of the world effectively filters out the relationship you have to your own experiences. In terms of the themes of this book, filtering the world through a secondhand connectedness makes it impossible for any of us to hear the song of the African's "persistent and collective intelligence." Might it be better to seek out that song than anything you find on the hit parade?

What does it mean, "a persistent and collective intelligence"? Regard the forum of Iroquois tribal government. Whenever the Iroquois held a council meeting, they first spoke an invocation:

> In our every deliberation we must consider the impact of our decisions on the next seven generations.

Thereafter, any vote among the living council members also included an equal vote for the needs and dignity of those who would live 150 to 200 years in the future.

The pre-conquest Iroquois would have agreed with Marshall McLuhan when he wrote that the medium is the message. In this case, the generational format of their council essentially defined a long-term relationship to the land. The rights of future generations never became an issue of policy because they were, instead, the actual context of policy. Conservation was the context upon which their government, culture, and community was built.

Thomas Jefferson was said to have drawn much inspiration from the structure of Iroquois democracy in the process of blueprinting our American system of government. It makes one wonder about things that might have been. How would our lives be different today if Jefferson had included the rights of the seventh generation in the U.S. Bill of Rights? Imagine if some far-sighted politician had voted against the intercontinental railroad in 1840, or against the national highway system in 1910. We might not have a greenhouse effect looming over our lives.

Our post-modern culture plods through the environment like one of its giant earthmovers, ill-equipped to base environmental policy upon anything beyond the needs of the current fiscal year. We have gone so far as to teach our children that their intuitive recognition of a long-forgotten interconnectedness with nature is a naïve fancy. We educate it out of them and simultaneously fill the void with an objective methodology that defines our current relationship to the land. But this way of perceiving nature transmogrifies the community of life into a supermarket of information, most elegantly depicted as facts and statistical models. There it resides under the supervision of those who understand the mathematical models, under the control of those who sponsor the biologist.

If the medium is the message, it seems fitting to redefine the science of ecology as this tongue twister: the lingual and linear translation of nature's non-lingual and non-linear holistic processes. In other words, whereas ecology should mean that we are what we live inside of, it actually means we are what we live outside of.

Listen to Luther. He seems to be telling us that we have hired the wrong bunch of people for the job at hand. Any such starry-eyed study of "the all and the everything", as George Gurdjieff involuntarily defined ecology, seems far better served by philosophers, shamans, and artists. These dreamers have always possessed a methodology capable of describing the whole because they do not try to stand outside of it. Permit me to put on a straight face and suggest that the culture's relationship to nature will improve dramatically on the day that Senate subcommittees concerning land use start consulting with musicians and deep ecologists as a matter of course, elevating them to the critical position currently occupied by scientists and politicians: that of defining and explaining nature for the rest of us.

So I state the central nagging argument of this book.

## I Search, You Search, We All Search for ICERC

That nagging feeling demands our full attention. As it pertains to dolphins, the world does not need another swim program, another dolphin-assisted therapy program. Even whale watching is suspect. Granted, human beings control the fate of wild animals everywhere on the planet, and whale and dolphin watching potentially serves the cause of preservation by educating people about the beauty and intellect of cetaceans. What we are willing to pay money to watch, we are likely to take a stand to protect. It does work that way sometimes.

Other times, too many people needy of a whale or dolphin "hit" create a local economy of entrepreneurs whose natural motivation is economic. The whales' welfare is important, but not primary. Because the potential for both harm and preservation exist simultaneously—sometimes in the same place at the same time—the modern phenomenon of whale watching may be best understood as the midpoint of a line with industrial whaling on one end and leaving the whales alone on the other end. This latter is best understood as the way we relate to robins, cows, and trees.

The metaphor of the line provides a handy indicator to help us understand when to promote whale watching and when to denounce it. On San Juan Island, Washington, where I live, I have been denouncing orca watching for some years now. The industry promotes itself as preservation-minded and strictly regulated, when in fact the regulations have no teeth. On any day between June and September, fifty or more noisy boats may be surrounding the orcas for ten or more hours. Common sense suggests that impeding a whale's movement and drowning out its ability to communicate has to affect the orca's general health.

On the other hand, in Japan, where whaling still exists, whale watching provides a potent strategy for appreciating living whales. On the Russian White Sea, where there are few economic prospects, a fledgling eco-tour economy has effectively abrogated a proposal to permit the hunting of beluga whales, either commercially or by sportsmen.

The time has come for us to either give something back to the cetaceans, or simply leave them alone. In just this spirit, there is in Australia a small group of dolphin and whale aficionados who collectively call themselves the International Cetacean Education and Research Center. Because it is such a long name, the organization has adopted the acronym ICERC (pronounced I Search). Over the years ICERC has sponsored several dolphin and whale conferences in Australia, Hawaii, Japan, and Europe.

During a conference held in a resort hotel in Hawaii, it dawned on me that the majority of the presentations differed radically from the usual brand of expert-oriented oceanographic research. This conference put far more emphasis on the then radical idea of developing relations with cetaceans rather than collecting data about them. What was being presented here was not so much RESEARCH (with all its big science objectivist baggage), but rather, "isearch" (spelled ICERC), a made-up word that leans heavily toward the experiential sunny side of life, representing a worldview in the process of being born, the realm of radical educators, artists, political visionaries, and metaphysicians of various stripes. Not only did the organization pull off the most fascinating dolphin and whale conference I'd ever participated in, but their actual name offered a savvy description of the new paradigm represented.

The key element distinguishing this conference from other cetacean events was the presence of several Japanese conferees seeking strategic guidance to provide a transformation in the Japanese cultural relationship to whales and dolphins. I learned from them that although much Western activist money has been spent over the years attempting to curtail Japanese whaling, the very method of foreign protest is not effective in Japan, with the result that all those environmental dollars had the cumulative effect among the Japanese public of creating sympathy for the embattled whale industry. To put an end to whaling, the Japanese people first had learn to perceive whales as valuable in their own right, instead of as resources providing jobs for a rather small industry.

The only Japanese that attended other international gatherings were angry politicians and arrogant industrialists presenting essentially unethical prerogatives to further impose Japan's much-maligned whaling industry on the world environmental stage. The International Whaling Commission's annual meeting, for instance, was always a venue for classic confrontation.

As a believer in the inherent rights of cetaceans, I regard the IWC stalemate as metaphysical and perceptual. Japan's extinction-precipitating policy could be supported only within a culture whose religion (Buddhism) promotes reincarnation, where the issue of extinction does not possess the bite of finality that it holds within our linear, Christian-based culture. This metaphysical explanation is not meant to imply that Buddhism actually teaches anything of the sort, or that the Japanese are even religious. Rather, "reincarnation" exists as a kind of background hum within Japanese culture, in the same way that "love" is a background hum in our own.

In my naïveté, I once presented this metaphysical suspicion in the form of a scientific paper delivered at a meeting of the IWC in Washington, D.C. I was practically hooted off the podium by a very upset Japanese delegation. The moderator, a pro-whale scientist of some international distinction, demanded that I address my remarks to the pertinent issue on the floor. I turned to him to reply that I was not being "impertinent," because the discussion on the floor focused on the reasons the Japanese representatives would not abide by a whaling moratorium recommended by nearly all the rest of the members. My comments caused dismay, not because they strayed from the subject at hand, but because they called into question the programmed structure of a male tribal gathering that would never succeed at ending whaling. Metaphysics was not on the program.

I spent most of two years working to stop the killing of dolphins by fishermen at Iki Island, Japan. During that time, I heard fishermen, bureaucrats, scientists, and laypeople graciously apologize to me for the continued slaughter, and then conclude that Japan must soon stop killing whales and dolphins. I was no disinterested party, but a cetacean activist in the land of the anti-whale. I eventually realized that these Japanese acquaintances were simply telling me what they thought I wanted to hear. But they weren't lying. They were, instead, being courteous, which is a cornerstone of Japanese social interaction. Regard it as a cultural expression of form over content.

When I questioned these apologists a bit deeper, they acknowledged that a majority of the Japanese people stood in favor of continued whaling. A few confessed to me that as long as Westerners criticized Japanese marine mammal policies, the government would never quit. Never. Within Japan, criticism evokes the strong reaction known as saving face, which comprises a more important ethic than the extinction of whale species. For this reason, the Japanese government has always argued that the so-called "problem" of industrial whaling is actually a problem of how to contain foreign criticism.

Unfortunately, most of the Western environmental groups who have worked so hard in the past to change the Japanese relationship to cetaceans have focused on being critical, often of the male tribe of whaling industrialists and their misbegotten rules. Activists mistakenly believe that an appeal to logic—exposing the political, economic, and environmental corruption involved—must inevitably stop whaling. They have followed this argument for some years to the exclusion of the personal, the intuitive, and the experiential relationship. The strategy has failed utterly because it seeks to impose a new paradigm on the Japanese from the outside in, and from the top down.

Given this very cursory examination of a complex problem, one may well ask how we get the Japanese out of whaling? If you had asked me that question before ICERC, I would have answered using the language of pathology: that the Japanese are locked in a culture-abetted addiction denial, an obsessive/compulsive neurosis not unlike our cultural addiction to fossil fuels. Whaling is a form of denial; Japanese leaders do not wish to hear any worthwhile point of view other than their own. Likewise, the industry's reliance on scientific reductionism, *reductio ad absurdum,* and legal subterfuge strangles the effectiveness of the IWC.

Healing the Japanese xenophobia that nurtures whaling can occur only by expanding the context of international discussions to include concepts beyond the legal, the scientific, and the economic: concepts like the sacred. If so, how do we convince the Japanese that a moratorium on whaling does not place their way of life at risk? Where is the positive strategy that supports both Japanese culture and the protection of whales?

Before ICERC, I would have answered that there is no positive strategy, and those of us against whaling need to put the Japanese more at risk by focusing our resources more intently on making the international criticism against killing whales a greater threat to Japanese stability than anything the whalers could counteract. Because of the Japanese's historic inability to even hear the terms of this debate, if they stop whaling, it will be because of a worsening international image problem and not because of any change of heart.

The Japanese ICERC organization offers an alternative vision that is neither scientific, educational, nor political. It is, rather, cultural, offering an experiential concept I choose to call dolphin appreciation. This enters the society through enhanced media images about whale watching, swimming with dolphins, deep ecology, and even the modern mystery schools that try to explain what the large-brained cetaceans might be thinking about.

I explored some of these ideas twenty years ago in *Dolphin Dreamtime.* The writing had its effect and even a heyday; it was translated and published in several languages, including Japanese. Then, as the publishing biz goes, one at a time publishers started dropping it from their lists. That is, except in Japan, where the book is still in circulation and continues to sell. This new epilogue to a new edition insists I acknowledge the book's impact within Japan.

## Get Connected

There is a joke about an anthropologist visiting San Bushmen to study animism, who could not find one person in the tribe who called the religion "animism." In fact, they did not have any word in their language for "religion." Or wilderness. Or nature.

For that matter, consider what Luther Standing Bear would make of his Native American spiritual connection to nature being called "Dreamtime" or "animism" or "religion." Or regard this possibility, voiced by anthropologist Paul Reisman:

> Our social sciences generally treat the culture and knowledge of other peoples as forms and structures necessary for human life that those people have developed and imposed upon a reality that we know of, or at least our scientists know of, better than they do. We can therefore study those forms in relation to "reality" and measure how well or ill they are adapted to it. In their studies of the cultures of other people, even those anthropologists who sincerely love the people they study almost never think that they are learning something about the way the world really is. Rather they conceive of themselves as finding out what other people's conceptions of the world are.

Another long-dead Indian chief, Seattle of the Duwamish tribe, has recently emerged as one of the spokesmen for these "other people" who shared a deep interconnectedness with nature.

> All things are connected like the blood which unites one family. Whatever befalls the Earth befalls the sons of the Earth.

Whether Seattle actually said such a thing, or whether it exists only in the film version of his life, the sentiment remains viable. Old Chief Seattle did not warn that people living 150 years in the future had better learn to carve demon masks and eat camas roots and dance around with feathers stuck to their behinds in order to survive. It seems a mean-spirited distraction to distort what is actually a savvy relationship to nature by belittling it as an animist philosophy in competition with our post-modern Zeitgeist. As Reisman suggests, wisdom is not a game of basketball with tall moderns battling short primitives. To place the Dreamtime into the cubicle of animism and then label it a digression that can never be adopted generally is like saying that perception and gut reactions can never be adopted generally.

The environmental crisis begins and ends inside each of us. Regard it as a crisis in human perception. How we live is what we perceive. Language informs perception. And artists are every culture's arbiters of perception. Once you see it that way, you begin to think a little more creatively about how to alter your own relationship to the planet. We get connected when we act connected.

Indigenous art attempted to preserve and animate the unity that exists between humans and nature. Consider the aboriginal Dreamtime concept known as a songline: a physical trail of chants sung at specific locations that serve to animate the natural features of the ecosystem. Regard a songline as a kind of geographical music score, a map for the ears that explains the relations between humans and nature as well as the resultant responsibilities. Sing the right song and you may locate water in a desert you have never visited before. Significantly, the aborigines do not possess a concept of ownership like our own. Land, spirit, and self are inseparable. The earth can no more be sold than can one's soul. So the songline serves as a kind of non-possessive aesthetic relationship to place.

Artists have always served culture as arbiters of perception, and thus could be doing much more to alter the fact that the environmental crisis is a crisis in human perception. We need a new aesthetic of natural interconnectedness.

In fact, it is evolving from several contemporary aesthetic forms including earth art, soundscaping, vision questing, biopoesis, and interspecies music. One of the best-known examples of environmental art was Christo's *Running Fence,* which wound its way across twenty-four miles of California pastureland. The fence balanced natural landscape with cultural landscape—a fence that ran right through fences, preventing the eye from stopping at the usual property lines. And as the fence exposed the natural landscape under the cultural overlay of property borders, it also exposed the prejudices, foibles, humor, and humorlessness of American property law, confronting the mystique of private ownership, rational science, and governmental authority as they relate to a sense of place. When the event of the fence had run its course, the material comprising the fence was dissembled. Postholes were filled in; fabric was cut into squares, framed, and sold as *objets d'art.* Profits paid for the event. Besides the photos, nothing else remained.

## And in the End . . .

The seventh generation silently scrutinizes our growing inability to penetrate the armor of our disconnectedness. And despite the occasional promotion of their situation that derives from igniting the imagination of a few prophetic scouts like the Iroquois lawgivers, they seem quite incapable of doing much of anything on their own behalf. They remain seated along the temporal sidelines, holding their collective breath, waiting in anticipation of how our future is going to transmute into their present. They fall down on their unborn knees before us, pleading for us to undertake the monumental task of transforming, not just issues, but culture itself. Who hears them?

The Australian aborigines certainly do. Although clearly, we need to incorporate indigenous ways of perceiving into any general recipe for survival, let it also be noted that we cannot hope to ask anyone living in the modern world to give up their steel tools, their underwear, and possibly their computers. Murray Bookchin, for one, chides what he considers to be an eco-lala sentiment of finding solutions that demand a regression back to the primitive lifestyle. It is a "holism [that] evaporates into a mystical sigh." He offers this practical advice:

> Indeed, there is a level at which our consciousness must be neither poetry nor science, but a transcendence of both into a new realm of theory and

practice, an artfulness that combines fancy with reason, imagination with logic, vision with technique. . . . Poetry and imagination must be integrated with science and technology, for we have evolved beyond an innocence that can be nourished exclusively by myths and dreams.

# References

PREFACE

Page ix Douglas Baglin, *People of the Dreamtime.* Walker/Weatherhill, New York, 1970.

CHAPTER 1

Page 6 Barry Commoner, *The Closing Circle.* Jonathan Cape, London, 1972; Knopf, New York, 1971.

Page 12 Alan Watts, *Tao, the Watercourse Way.* Jonathan Cape, London, 1976; Pantheon, New York, 1975.

CHAPTER 2

Page 20 Gary Snyder, "Wild, Sacred, Good Land," in *The Schumacher Lectures, Volume II.* Edited by Satish Kumar. Blond & Briggs, London, 1984; and in *CoEvolution Quarterly,* Fall 1983.

Page 24 Tom Robbins, *Even Cowgirls Get the Blues.* Bantam, New York, 1990; Corgi, London, 1977.

CHAPTER 3

Page 35 Henry Beston, *The Outermost House.* Doubleday, Garden City, New York, 1930.

Page 40 Yamamoto Tsunetomo, *Hagakura.* Translated by William Scott Wilson. Kodansha International, Tokyo, 1979.

Page 46 Richard Ellis, *Dolphins and Porpoises.* Knopf, New York, 1982.

CHAPTER 5

Page 78 Seyyed Nasr, "Progress and Evolution," in *Parabola,* Vol. VI, May 1981.

Page 78 Joan Halifax, *Shaman, The Wounded Healer.* Crossroads, New York, 1982.

Page 78 Attributed to the "*Magic Words,*" Nalugiaq Eskimo "poem."

Page 80 Max Planck, *Scientific Autobiography and Other Papers.* Translated by F. Gaynor. Greenwood, New York, 1949.

CHAPTER 6

Page 86 Doug Boyd, *Rolling Thunder.* Random House, New York, 1974.

Page 89 Ibid.

CHAPTER 8

Page 118 From the daily log of the Orca Research Center, Alert Bay, Canada. Written by caretaker Jim O'Donnell.

Page 120 Thomas Kuhn, *The Structure of Scientific Revolutions.* University of Chicago Press, 1970.

Page 125 Sir Richard Burton, *Vikram and the Vampire*. Dover, New York, 1969.

CHAPTER 9

Page 138 Wade Doak, "Anecdotal Reporting of Human/Dolphin Interaction." Paper presented at the Whales Alive Conference, Boston, June 1983.

Page 139 Commoner, op. cit.

Page 140 Rupert Sheldrake, *A New Science of Life*. Muller, Blond & White, London, 1985; Tarcher, Los Angeles, 1981.

Page 144 Edward T. Hall, *Beyond Culture*. Anchor Doubleday, Garden City, New York, 1976.

Page 145 Ted Mooney, *Easy Travel to Other Planets*. Farrar, Straus & Giroux, New York, 1981.

EPILOGUE

Page 163 *Touch the Earth*. Compiled by T.C. McLuhan. Touchstone Books, New York, 1971.

Page 165 John Berger, *About Looking*. Pantheon, New York, 1980.

Page 165 David Dunn, "Nature, Sound Art and the Sacred," from *The Book of Music and Nature*. Edited by David Rothenberg and Marta Ulvaues. Wesleyan University Press, Middletown, CT, 2001.

Page 166 Great Law of the Haudenosaunee: the Six Nation Iroquois Confederacy.

Page 171 *Seeing Castenada*. Edited by Daniel Noel. G.P. Putnam's Sons, New York, 1976; quoted in Morris Berman, *The Reenchantment of the World*, Bantam, New York, 1984.

Page 171 Chief Seattle, on the occasion of surrendering his land, upon which the city of Seattle now stands, in *Touch the Earth*, op. cit.

Page 172 Peter Warshall, "Christo's Running Fence," in *CoEvolution Quarterly*, Winter 1976/77.

Page 173 Murray Bookchin, "The Concept of Social Equality," in *CoEvolution Quarterly*, Winter 1981.

# Bibliography

**Crail, Ted.** *Apetalk and Whalespeak.* Tarcher, Los Angeles, 1981.

**Dillard, Annie.** *Pilgrim at Tinker Creek.* Harper's Magazine Press, New York, 1974; Jonathan Cape, London, 1975.

**Doak, Wade.** *Dolphin, Dolphin.* Sheridan House, New York, 1982.

**Hofstadter, Douglas.** *Gödel, Escher, Bach.* Vintage Books, New York, 1980.

**Khan, Inyat.** *Music.* Sufi Press, India, 1962.

**Leopold, Aldo.** *The Sand County Almanac.* Oxford University Press, New York, 1949.

**Lilly, John.** *Lilly on Dolphins.* Anchor Doubleday, Garden City, New York, 1975.

**Lovelock, James.** *Gaia.* Oxford University Press, New York, 1979.

**Robson, Frank.** *Thinking Dolphins, Talking Whales.* Reed, Wellington, New Zealand, 1976.

**Russell, Peter.** *The Global Brain.* Tarcher, Los Angeles, 1983.

**Tompkins, Peter, and Christopher Bird.** *The Secret Life of Plants.* Harper & Row, New York, 1973; Allen Lane, New York, 1974.

**Van der Post, Laurens.** *The Lost World of the Kalahari.* Morrow, New York, 1958; Hogarth, London, 1961.

# Index

## P



## R



## S



## T



## V



## W



## Z

## ABOUT THE AUTHOR

Jim Nollman was born in Boston in 1947 and graduated from Tufts University in 1969. He has been a composer of music for theater, an internationally distinguished conceptual artist, and an environmental activist. In 1973, he was commissioned to compose a Thanksgiving Day radio piece for a U.S. national network, and recorded himself singing children's songs with three hundred turkeys. He has recorded interspecies music with wolves, desert rats, deer, elk, whales, and dolphins. He directed one of Greenpeace's first overseas projects, at Iki Island, Japan, where fishermen were slaughtering dolphins to compensate for human overfishing.

He is the founder of Interspecies, which sponsors research on communicating with animals through music and art, and promotes a model for communion between species. Interspecies's best-known field project is a twenty-five-year study using live music to interact with the wild orcas who inhabit the west coast of Canada. Nollman is currently directing a project in Arctic Russia to protect the last beluga whales in Europe, and is learning how to communicate with these whales. For more information about this ongoing work, and to hear many examples of music between species, visit the Interspecies website at *interspecies.com.*

Jim Nollman is the author of five books, and his essays are anthologized in several collections of nature writing. He is contributing editor of the largest whale site on the Internet, with 10,000 visitors a day. He lives on San Juan Island in the northwest corner of the United States with his wife, Katy, and daughters, Claire and Sasha.

Sentient Publications, LLC publishes books on cultural creativity, experimental education, transformative spirituality, holistic health, new science, and ecology, approached from an integral perspective. Our authors are intensely interested in exploring the nature of life from fresh perspectives, addressing life's great questions, and fostering the full expression of the human potential. Sentient Publication's books arise from the spirit of inquiry and the richness of the inherent dialogue between writer and reader.

We are very interested in hearing from our readers. To direct suggestions or comments to us, or to be added to our mailing list, please contact:

SENTIENT PUBLICATIONS, LLC
1113 Spruce Street
Boulder, CO 80302
303-443-2188
contact@sentientpublications.com
www.sentientpublications.com